Elementary Quantum Mechanics

MORTEN SCHARFF

Niels Bohr Institute
University of Copenhagen
Denmark

WILEY—INTERSCIENCE

A division of JOHN WILEY & SONS LTD.

London New York Sydney Toronto

First published under Morten Scharff, *Elementær Kvantemekanik*
by Akademisk Forlag, Copenhagen

Library of Congress catalog card No. 72-75365

SBN 471 75700 4

Made and printed in Great Britain by
Willmer Brothers Limited, Birkenhead

Preface

Morten Scharff died in 1961 at the age of 34. In the years before his death he gave a new undergraduate course in quantum mechanics at the Niels Bohr Institute in Copenhagen. His lecture notes, with a few later additions, were published in 1963 as a textbook, now used extensively at Danish universities. The special approach in Scharff's course makes it different from most textbooks on the subject. For this reason we believe that it is desirable to make the course accessible to a wider public.

Morten Scharff aimed at giving the students a strong impression of the physical contents of wave mechanics without entering too far on the formal aspects of the theory. He hoped that then the students might more easily cope with actual problems encountered in physics, problems outside the smoother texture of physics courses. In order to achieve this end he laid particular emphasis, for instance, on the comparison with classical mechanics by extensive use of wave packets. He often sacrificed an elegant exposition in favour of a more pedestrian one, founded directly on basic ideas of quantum mechanics. The same purpose governed his choice of examples of atomic physics, where the application of quantum mechanics is illustrated by simplified, but relevant, models of crucial phenomena. By this selection of the material Morten Scharff paid attention also to the needs of those students who ask for an understanding of the salient features of quantum mechanics without necessarily aiming at becoming specialists.

One does not need extensive knowledge in mathematics or physics in order to be able to study the present book. Still, it is desirable to have phenomenological insight in atomic and nuclear physics, as well as some knowledge of classical mechanics and its simplest applications.

The English version is essentially a direct translation of the original manuscript by Morten Scharff, with some additions. Major additions are 2.6, 3.4, 3.5, 3.8 and Chapter 5; we hope that they are in accord with the general trend of the original course. We wish to thank Mrs. A. M. Winther and Mrs. L. Wilkinson for the translation.

Jens Lindhard Aage Winther
Aarhus, March 1968.

Contents

1

Basic Principles of Wave Mechanics

1.1 INTRODUCTION

The foundations of non-relativistic wave mechanics, the branch of quantum theory to be treated here, were laid down by Schrödinger in 1926. Shortly before this, Heisenberg had formulated a quantum mechanics based on matrix algebra. His theory was soon shown to be equivalent to Schrödinger's, but for many practical purposes wave mechanics proved easier to work with than matrix mechanics.

In a few years this new quantum mechanics was developed to a high degree of perfection by a large number of physicists, among whom Bohr, Born, and Heisenberg were particularly active in formulating a consistent physical interpretation of the mathematical apparatus of the theory. In 1927 wave mechanics was generalized by Pauli to apply to particles with spin, and in 1928 Dirac proposed the relativistic wave equation for the electron.

The new quantum mechanics dating from 1925–26 replaced the so-called old quantum mechanics which had its roots in Planck's introduction, at the turn of the century, of the concept of quanta in physics. In order to explain the frequency distribution in heat radiation from solid bodies, Planck was obliged to assume that radiation was transmitted in quantum units, each containing the energy

$$E = h\nu, \tag{1.1.1}$$

ν being the frequency of the radiation emitted, while h is Planck's universal constant, having the dimension of an action (energy × time), and magnitude

$$h = 6{\cdot}6 \times 10^{-27} \text{ erg sec.} \tag{1.1.2}$$

This quantum description of electromagnetic radiation, supported by Einstein's interpretation of the photo-electric effect (1905), was utilized by Bohr (1913) in a theory capable of describing the line spectrum of hydrogen on the basis of Bohr's famous postulates. This formed the basis of the 'old' quantum mechanics which was then developed by Bohr, Sommerfeld, and others to explain the discrete energy states of the atoms and to give a qualitative description of atomic spectra and of the periodic system of chemical elements.

The old quantum mechanics* was essentially a classical mechanics, subject to restraints, the so-called quantum conditions, which select certain of the infinitely many possible classical orbits as the only permissible ones. The rules of quantization could be applied only to systems classically performing periodic movements such as, for example, classical elliptical orbits of an electron in a Coulomb field, or harmonic vibrations of atoms about their mean position within a molecule. On the other hand, the theory could not be made to apply to systems which change in time, as in radioactive decay or collisions between atomic particles. Hence the old quantum mechanics could by no means be considered a final and consistent theory.

Before proceeding to the treatment of wave mechanics and its results, it is worth mentioning how it was possible to link wave concepts with material particles. In 1923 the Compton effect was discovered; it showed that in the case of scattering of short-wave electromagnetic radiation by electrons, an electromagnetic wave should be considered as a stream of particles, each having the energy $E = h\nu$ and a momentum in the direction of propagation of the wave of magnitude

$$p = h/\lambda, \tag{1.1.3}$$

h being Planck's constant and λ the wavelength. In this case the radiation quanta are particles of mass zero (photons).† Hence in certain experiments electromagnetic radiation should be regarded as a stream of particles, whereas, on the other hand, the pronounced

* The old quantum mechanics is discussed in detail in S. Tomonaga, *Quantum Mechanics*, North Holland, Amsterdam, 1963.

† ν and λ for electromagnetic radiation being connected by the relation $c = \nu\lambda$, from (1.1.1) and (1.1.3) we obtain $E = pc$.

interference effects in the case of the passage of X-rays through crystals (Bragg and Laue diffraction) show that here the electromagnetic radiation occurs in the form of waves.

In 1924 de Broglie proposed, as a hypothesis, that the above wave–particle dualism in the description of electromagnetic radiation should be extended to apply also to material particles, in such a way that it should be possible to associate with a particle a wave of frequency and wavelength determined by, respectively, the energy and momentum of the particle as in (1.1.1) and (1.1.3). This hypothesis was supported experimentally in 1927, when Davisson and Germer, by scattering electrons of known momentum on a nickel crystal, found that the electrons emerged in precisely the directions where one would expect to find intensity maxima for the diffraction of X-ray radiation of a wavelength depending on the momentum of the electrons as in (1.1.3).

This suggests that there is an analogy between wave mechanics for material particles and Maxwell's wave theory for electromagnetic radiation. Although the analogy can be pushed too far (for one thing, photons have zero mass and cannot therefore be described in non-relativistic terms), it may serve to hint at the way in which wave mechanics differs from classical mechanics. The analogy will be used in the following, where an attempt is made to present some plausibility arguments for the basis of wave mechanics.

Problem

1.1.1 Calculate the de Broglie wavelength for electrons with an energy of 100 eV, as well as for particles with mass 1 g and velocity 1 m/sec.

1.2 THE SCHRÖDINGER EQUATION FOR A PARTICLE

Quantum mechanics is a mathematical formalism, which, on the basis of knowledge of the state of a system, at a given time, allows the calculation of the state of the system at a later time, so that the result of experiments can be predicted. In quantum mechanics, the state of a particle is described by a wave function. At any given moment, the wave function describes a certain field in space, associated with the immediate state of the particle. The wave function develops in time in a manner depending on the environment of the particle, and the state of the particle at a later time is then usually described by another field in space.

In order to use this type of formalism it is necessary to translate experimental knowledge of the physical state of particles into mathematical properties of the wave function and vice versa, and one must also be able to calculate the development in time of a given wave function. For the latter purpose we use a differential equation, the Schrödinger equation, which governs the behaviour of the wave field in time and space and hence occupies a position analogous to that of Maxwell's equations within the electromagnetic field theory.

As to the association of a wave function with a given physical situation, there is a decisive difference from the electromagnetic analogy. In fact, while the electromagnetic fields, the electric and magnetic field strengths, are directly measurable physical quantities, the quantum mechanical wave function is an auxiliary concept with less direct connection with the experiments.

It is possible to present quantum mechanics in an axiomatic form by postulating the Schrödinger equation and a set of rules for the physical interpretation of the wave function. However, in the following we shall use a less systematic approach, introducing the different aspects of the formalism by degrees, and trying to render the assumptions plausible.

We shall take Davisson and Germer's experiment as our starting point, and we shall further place the rather natural demand on our wave description that in the limit—at least for macroscopic phenomena—it must agree with Newtonian mechanics. For a detailed description of the Davisson and Germer experiment, and of other diffraction experiments with material particles (electrons as well as atoms and molecules), see, for example, H. Semat, *Introduction to Atomic Physics*.

The essence of these experiments may be summed up in the following way. When a beam of atomic particles of well-defined momentum p hits a crystal, scattered particles are observed mainly in certain directions; these are distributed in such a way that when the particles are registered on a photographic plate, it shows an interference pattern perfectly analogous to the one which would result if, in the experimental arrangement, the incident beam of particles were replaced by monochromatic X-rays of wavelength $\lambda = h/|\mathbf{p}|$. This relation between the momentum of the particles and the wavelength determined by the interference picture is particularly accurate only for fast particles. We shall return to the significance of the deviations; at the moment we ignore such elaborations.

Having made this reservation, one may 'explain' the electron diffraction experiments by formally associating the beam of particles with a plane harmonic wave train proceeding in the direction of the beam of particles, and having a wavelength which depends on the momentum of the particles in the above manner. There is, of course, no necessity for the associated wave to be of an electromagnetic nature; but it must be possible to imagine that it is diffracted in crystals. The only necessary condition for this is that the incident wave train be in some way disturbed by the strong force fields near the atoms of the crystal lattice so as to give rise to scattered waves oscillating at the same frequency as the incident wave. Furthermore, since interference must be possible, we must assume the validity of the superposition principle for the associated waves, in the same manner as for electromagnetic fields. These two assumptions are the only ones needed to arrive at the Bragg condition for the directions in which the scattered waves from a large number of lattice points are added in phase, and therefore combined to a scattered wave of large amplitude.

The amplitude of the scattered waves hitting the various points of the photographic plate must in one way or another be interpreted as a measure of the flow of electrons in this region of space, in the same way as the intensity of the electromagnetic waves determines the flow of photons in the X-ray diffraction experiment. However, in our preliminary considerations we need not worry about the quantitative relations between the associated wave field and the distribution of the electrons in space. It is sufficient to assume that the electrons occur mainly where there are waves of large amplitudes, whereas no electrons are found where the waves cancel each other.

Before continuing the discussion of the associated waves we shall consider wave fields in general. A wave function describing a plane harmonic wave may be expressed in a number of ways, e.g. by either of the following three representations

$$\psi(\mathbf{r},t) = \begin{cases} A \sin(\mathbf{k}\cdot\mathbf{r}-\omega t) & \text{(a)} \\ A \cos(\mathbf{k}\cdot\mathbf{r}-\omega t) & \text{(b)} \\ A \exp[\mathrm{i}(\mathbf{k}\cdot\mathbf{r}-\omega t)], & \text{(c)} \end{cases} \qquad (1.2.1)$$

where A is a constant (the amplitude).

The above field quantity, ψ, might be a component of the electric field strength. In this case the complex formulation (1.2.1c) is only used for convenience, the field strength being represented by the real part.

In the expressions (1.2.1) it is seen that the phase, $\mathbf{k}\cdot\mathbf{r}-\omega t$, at a given time is constant on any plane normal to $\mathbf{k}$, since such planes are represented by $\mathbf{k}\cdot\mathbf{r}$ = constant. The vector $\mathbf{k}$, which is called the wave vector, is hence normal to the wave plane. The phase difference between two wave planes at a mutual distance λ is $|\mathbf{k}|\lambda$. If λ is to represent the wavelength, the phase must increase by 2π over this distance, and hence the magnitude of the wave vector (the wave number)* must be given by

$$k = |\mathbf{k}| = \frac{2\pi}{\lambda}. \qquad (1.2.2)$$

If the x axis is chosen in the direction of $\mathbf{k}$, the phase may also be written in the form $kx-\omega t$, or $2\pi(x/\lambda-\nu t)$, where $\nu = \omega/2\pi$.

Over a period of time, ψ oscillates at a fixed point in space with the angular frequency ω. At a fixed point with abscissa x, the phase decreases over a short period of time δt by the amount $\omega\delta t$, but this is compensated for if we simultaneously increase x by the quantity $\delta x = (\omega/k)\,\delta t$. Thus the wave planes with a given phase proceed in the direction of $\mathbf{k}$ with the phase velocity (see Fig. 1.1)

$$v_p = \frac{\omega}{k}. \qquad (1.2.3)$$

It is possible to observe neither the angular frequency ω nor the phase velocity in diffraction experiments which determine only the wave vector $\mathbf{k}$. For electromagnetic waves the frequency can be found from the wavelength since the waves comply with

* Often, especially in older literature, the wave number is defined as $k = 1/\lambda$, the quantity $2\pi/\lambda$ then being called the circular wave number.

the Maxwell equations, leading to a wave equation of the type

$$\Delta\psi - \frac{1}{c^2}\frac{\partial^2\psi}{\partial t^2} = 0, \tag{1.2.4}$$

where $\Delta = \partial^2/\partial x^2 + \partial^2/\partial y^2 + \partial^2/\partial z^2$ is the Laplace operator, while c is the velocity of light.

For wave functions of type (1.2.1), calculations yield $\Delta\psi = -k^2\psi$ and $\partial^2\psi/\partial t^2 = -\omega^2\psi$, i.e. (1.2.4) is satisfied by harmonic waves only if $c^2k^2 = \omega^2$, i.e. $c = \lambda\nu$. Since (1.2.4) is linear in ψ, it is satisfied by arbitrary linear combinations of functions which are themselves solutions, and in this way it is possible to form new solutions of (1.2.4) by the superposition of harmonic waves having the proper relation between ω and k.

In order to derive the wave equation for our beam of particles, we shall proceed in the opposite manner. From the demand for correspondence with the Newtonian mechanics we shall first find the relation between k and ω which must be obeyed by the superposition of the harmonic waves, and then see which differential equation is satisfied by these waves.

By the superposition of harmonic waves with different wavelengths it is possible to form a wave function which has large amplitudes only in a small region of space, where the waves are chosen to have the same phase, whereas they tend to cancel elsewhere.

At first we try to superpose waves in one dimension with wave numbers in a small interval between $k-\delta k$ and $k+\delta k$. Choosing

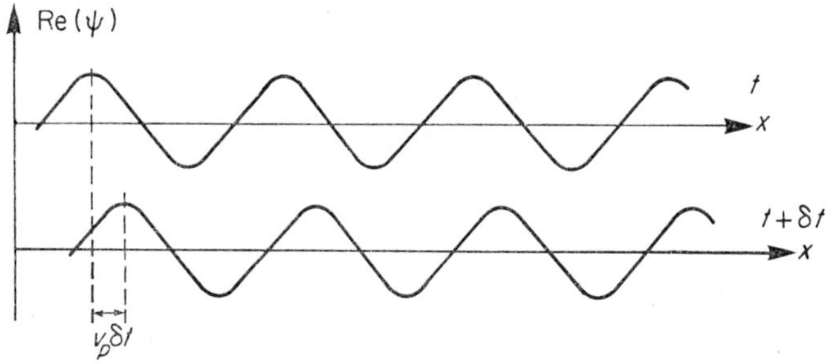

Fig. 1.1

a complex formulation we get for the resulting wave function at the time $t = 0$

$$\psi(x) = A \int_{k-\delta k}^{k+\delta k} \exp(ik'x)\, dk' = A\left[\frac{\exp(ik'x)}{ix}\right]_{k-\delta k}^{k+\delta k}$$

$$= A \exp(ikx)\,\frac{2\sin(\delta k \cdot x)}{x},$$

$\psi(x)$ represents a wave group, also called a wave packet, as indicated in Fig. 1.2.

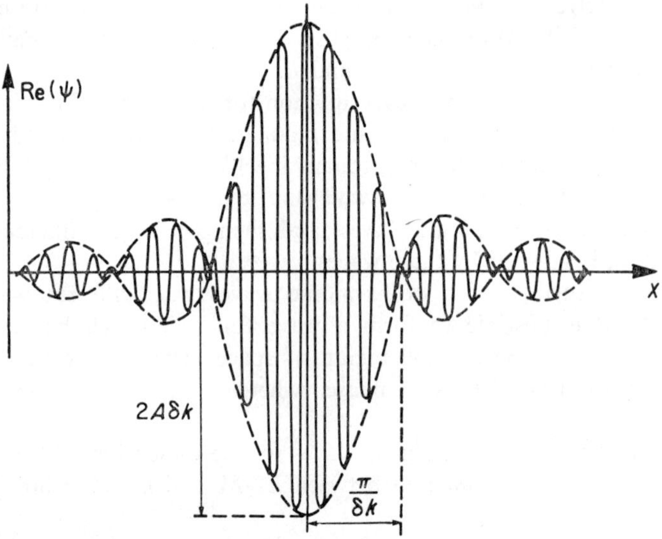

Fig. 1.2

It is seen that the wave function essentially vanishes outside a region whose width is inversely proportional to the width $2\delta k$ of the interval from which we have chosen our wave numbers. Here we have superposed harmonic waves with equal weight from the entire interval. By choosing different weight functions it is possible to change the shape of the wave packet, but its extent cannot be reduced below the order of magnitude of

$$\delta x \approx 1/\delta k, \tag{1.2.5}$$

where δk represents the width of the wave number interval which contributes materially to the wave packet. The relation (1.2.5) may be understood qualitatively by considering that harmonic waves which are in phase in the centre of the wave group, must have a difference in wave number of δk in order to get into counter-phase at the distance δx from the centre, and that consequently $\delta k\ \delta x \approx \pi$.

Similarly, one may construct a wave packet in space by superposition of harmonic waves with wave vectors $\mathbf{k}'$, which deviate slightly both in magnitude and direction from a fixed vector $\mathbf{k}$.

$$\psi(\mathbf{r},0) = A \int g(\mathbf{k}') \exp(\mathrm{i}\mathbf{k}' \cdot \mathbf{r})\, \mathrm{d}^3k', \tag{1.2.6}$$

where we integrate over all k' space with a weight factor $g(k')$, which is vanishingly small outside the domain

$$|k_x - k'_x| \lesssim \delta k,\ |k_y - k'_y| \lesssim \delta k \text{ and } |k_z - k'_z| \lesssim \delta k.$$

The linear extent of the wave group in space can then be of the order of magnitude of $1/\delta k$ in all directions.

The development in time of this type of wave packet is determined by the frequency of the harmonic waves incorporated in the wave packet, and it is necessary here to remember that ω is a function of the wave number. The wave packet (1.2.6) is at the time t then given by

$$\psi(\mathbf{r},t) = A \int g(\mathbf{k}') \exp(\mathrm{i}\mathbf{k}' \cdot \mathbf{r} - i\omega(\mathbf{k}')t)\, d^3k'. \tag{1.2.7}$$

We now exploit the fact that the wave packet includes only wave vectors $\mathbf{k}'$ very close to k, and that therefore ω in the integrand will deviate only slightly from the value $\omega(\mathbf{k})$. To first approximation we then find $\omega(\mathbf{k}') = \omega(\mathbf{k}) + \mathrm{grad}_k\omega \cdot (\mathbf{k}' - \mathbf{k})$, discarding terms of second and higher order in the quantities $k'_x - k_x$, $k'_y - k_y$, and $k'_z - k_z$.

Equation (1.2.7) may then be written in the form

$$\begin{aligned}\psi(\mathbf{r},t) &\approx A \exp[\mathrm{i}\mathbf{k} \cdot \mathbf{r} - \mathrm{i}\omega(\mathbf{k})t] \times \\ &\int g(\mathbf{k}') \exp[\mathrm{i}(\mathbf{k}' - \mathbf{k})(\mathbf{r} - t\, \mathrm{grad}_k\omega)] d^3k' \\ &= A \exp[\mathrm{i}\mathbf{k} \cdot \mathbf{r} - \mathrm{i}\omega(\mathbf{k})t]\ F(\mathbf{r} - t\, \mathrm{grad}_k\omega).\end{aligned} \tag{1.2.8}$$

In this approximation the integral in (1.2.8) is evidently a function of $\mathbf{r} - t\ \mathrm{grad}_k\omega$, and consequently the wave packet proceeds with the velocity

$$\mathbf{v}_g = \mathrm{grad}_k\omega. \tag{1.2.9}$$

This so-called group velocity for the waves is determined by the dependence of frequency on wave number, and it will therefore usually be different for wave packets of different 'mean wave vectors' **k**. Fig. 1.3 gives an idea of the movement of a one-dimensional wave packet. Since phase velocity usually differs from group velocity, we have to imagine that wave peaks are being displaced within the group, i.e. that new peaks develop continually, while others disappear.

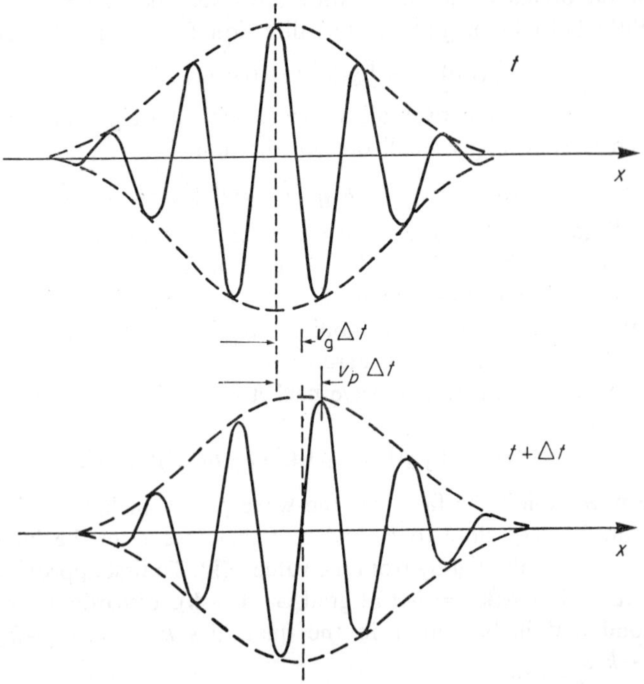

Fig. 1.3

It should be noted that the approximation leading to (1.2.9) is usually good only for a limited time, since the discarded terms in the phase are no longer insignificant compared to π when sufficient time has elapsed. Hence over long periods of time the wave packet may change its shape and become flattened [see problem 1.2.1].

For electromagnetic waves in vacuum we have $\omega = c\,k$, which implies that phase velocity and group velocity are equal. In this case the above approximation is exact for a one-dimensional wave group, which hence will be transmitted in unchanged form. Similar relations apply to sound waves, as may be concluded from the fact that music is heard without distortion at any distance from its source. However, in the case of surface waves of liquids there is a more complicated relationship between ω and k.

We now return to considering the waves which are to be associated with the incident beam of particles in the Davisson and Germer experiment. If we assume that the electrons have momentum $\mathbf{p}$, we have seen that we must assign to them a plane harmonic wave train with a wave vector in the direction of $\mathbf{p}$ of magnitude $k = 2\pi/\lambda = 2\pi p/h$. This may be written

$$\mathbf{k} = \frac{1}{\hbar}\mathbf{p}, \tag{1.2.10}$$

where we have introduced the commonly used abbreviation $\hbar = h/2\pi$. By superposition of such harmonic waves with wave vectors deviating only slightly from $\mathbf{k}$, and therefore, according to (1.2.10), all associated with particles of momenta virtually identical with $\mathbf{p}$, we can build up a wave function representing a wave packet. This wave function differs from zero only within a small domain of space, and it must therefore be possible to link it to a particle present in this domain and with a momentum $\approx \mathbf{p}$. If we desire to reduce the domain to linear dimensions of an order of magnitude a, it is necessary to superpose wave vectors with components in intervals of the order of magnitude $1/a$, which, according to (1.2.10), corresponds to a distribution of the components of the momentum within an interval of a minimum order of magnitude of $\hbar/a$. However, in the case of macroscopic particles, this 'haziness' in the momentum will be totally imperceptible. If, for example, we imagine a particle with a mass 1 g, and localized with the high accuracy of 10^{-4} cm, the construction of a wave packet of these dimensions will cause a lack of definition or spread in the velocity components of approximately $\hbar/am \approx 10^{-23}$ cm/sec.

In order to make our wave description compatible with classical mechanics, it is apparently necessary to demand that the above wave packet has the velocity $\mathbf{v} = \mathbf{p}/m$. We must therefore insist that this is the group velocity for waves with wave vector $\mathbf{k} = \mathbf{p}/\hbar$.

By using the expression (1.2.9) for the group velocity we then find

$$\mathrm{grad}_k\omega = \frac{\mathbf{p}}{m} = \frac{\hbar\mathbf{k}}{m},$$

which may immediately be integrated to give

$$\omega = \frac{\hbar\mathbf{k}^2}{2m} + C, \tag{1.2.11}$$

where C is an arbitrary constant.

Hence an arbitrariness is permissible in the absolute magnitude of the frequency (but not in its dependence on **k**), causing, according to (1.2.3), an arbitrariness in the magnitude of the phase velocity. This presents no difficulty, since in the case of associated waves the phase velocity and the frequency need not be observable quantities. This is connected with the fact that the wave function, contrary to the case of electromagnetic fields, is not accessible to direct measurements.

We can now construct a wave equation for our associated waves. In order to allow superposition of the solutions, the equation must be linear. Using (1.2.11), and for convenience putting $C = 0$, we can express the phase of the harmonic waves which are to be the solutions of the desired differential equation thus:

$$\varphi = \mathbf{k}\cdot\mathbf{r} - \frac{\hbar k^2}{2m}t. \tag{1.2.12}$$

We may then choose between the various possibilities for harmonic wave functions (1.2.1) which, with the phase (1.2.12), are

$$\psi(\mathbf{r},t) = \begin{cases} A\sin\varphi & \mathrm{I} \\ A\cos\varphi & \mathrm{II} \\ A\exp(\mathrm{i}\varphi) & \mathrm{III} \\ A\exp(-\mathrm{i}\varphi) & \mathrm{IV} \end{cases}$$

Guided by the form of the electromagnetic wave equation (1.2.4), we find by direct calculation that $\Delta\psi = -k^2\psi$ for all four wave functions. However, if we take the partial derivative of ψ with respect to time, we see immediately that forms III and IV distinguish themselves by the fact that already the first derivative is of the form: (constant)$\times k^2\psi$, which offers the possibility of a particularly simple wave equation. If wave function III is chosen, we find that $\partial\psi/\partial t = -\mathrm{i}(\hbar/2m)\,k^2\psi$, and we may then use the wave equation

$$-\frac{\hbar^2}{2m}\,\Delta\psi = \mathrm{i}\hbar\,\frac{\partial\psi}{\partial t}\,. \tag{1.2.13}$$

This differential equation is the Schrödinger equation for free particles of mass m. Its harmonic solutions are of the form

$$\psi(\mathbf{r},t) = A\exp\left[\mathrm{i}\left(\mathbf{k}\cdot\mathbf{r} - \frac{\hbar k^2}{2m}\,t\right)\right]. \tag{1.2.14}$$

The choice of the wave function III in preference to IV is merely a convention. The choice of IV would only cause a change of sign on one side of (1.2.13). On the other hand, wave functions of types I and II would require a differential equation of the fourth order in the space coordinates. The simple form (1.2.13) admittedly requires the use of complex wave functions, but this only means that we work with two real functions at once, the real and the imaginary part of ψ. Thus, when we say that the state of a particle is described by a wave function, then we actually mean a set of two real functions*. It may also be noted that the differential equation (1.2.13) is of only first order in time, and that consequently the dependence of the wave function on the space coordinates at a certain time determines $\psi(\mathbf{r},t)$ for all times.

Before we move on to a discussion of the interpretation of the wave function on the basis of the Schrödinger equation, we shall establish the Schrödinger equation for a particle in a conservative force field. Here we might extend our considerations concerning free particles, attempting to construct wave packets moving in the classical manner in macroscopic force fields, in order to arrive at the wave equation. However, in order to ease our progress, we will make a few more guesses.

We observed previously that in the case of the Davisson and Germer experiment the relation (1.2.10) between the momentum $\mathbf{p}$ of the incident electrons and the associated wave vector $\mathbf{k}$ holds strictly only for sufficiently fast electrons. In order to explain the interference picture with decreasing kinetic energies, $T = p^2/2m$,

---

* In relativistic quantum mechanics, wave functions with even more components are used.

for the electrons, it became necessary to introduce into the Bragg formula a slightly smaller wavelength than h/p, corresponding to a wave number slightly higher than $p/\hbar$. For nickel crystals the variation with T of the corrected wave number k' could be represented by a formula

$$k' = \frac{p}{\hbar}\sqrt{\frac{T+21 \text{ eV}}{T}}.$$

This result might still be interpreted in accordance with the de Broglie hypothesis, if only it were assumed that the electrons move with a slightly higher momentum

$$p' = p\sqrt{\frac{T+21 \text{ eV}}{T}}$$

within the crystal than outside. This type of effect might actually be expected, since energy is required to remove an electron from the crystal. Inside the crystal the electron is, on the average, affected equally in all directions by the surrounding atoms, but at the surface the actions add to an approximate jump in the potential for the electron.

By using $T = p^2/2m$, one finds that the above equation for p' is equivalent to

$$\frac{p'^2}{2m} - 21 \text{ eV} = \frac{p^2}{2m} = T, \tag{1.2.15}$$

which may be regarded simply as the energy integral for the electrons, assuming that the potential energy is $\Phi_0 = -21$ eV inside the crystal and $\Phi = 0$ outside.

In actual fact these considerations do not teach us anything new about our associated waves. The relation (1.2.15) merely determines how, for a given external energy T of the electrons, we may calculate the magnitude of the wave vector in (1.2.14), in order that this wave function may be associated with the electrons inside the crystal. However, the Davisson and Germer experiment also teaches us something about the wave function outside the crystal. It might be hoped that there we have a de Broglie wavelength corresponding to the electron momentum, p, outside the crystal. If this is the case, we have waves of different wavelengths outside and inside, and one may then expect refraction of a plane wave train which does not strike the surface at a right angle. This must, of course, be taken into consideration when comparing the external interference picture and the diffraction angles calculated on the basis of the Bragg conditions within the crystal. It was indeed found in the Davisson and Germer experiment that in order to obtain full agreement with the observed interference picture one must assume a refractive index

$$\frac{\sin \varphi}{\sin \varphi'} = \frac{\lambda}{\lambda'} \tag{1.2.16}$$

between crystal and vacuum.

Here φ and φ' are the angles between the normal of the crystal surface and the wave normals outside and inside the crystal respectively, while λ and λ' are the de Broglie wavelengths corresponding to the electron momenta in the two regions. An elementary consideration (see Fig. 1.4) shows that the refractive index is the

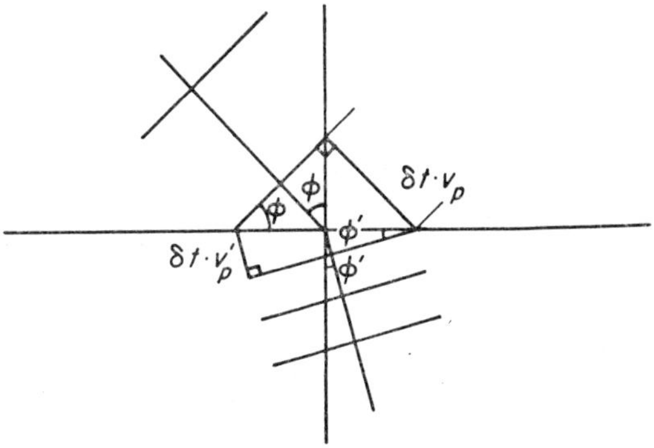

Fig. 1.4

ratio between the phase velocities of the two regions, so that

$$\frac{\sin \varphi}{\sin \varphi'} = \frac{v_p}{v_p'} .$$

From (1.2.3) and $\lambda/2\pi = 1/k$, one finds the refractive index (1.2.16) when one assumes identical frequencies in the two regions.

We have now learnt how to construct our associated wave function when we want to describe electrons moving with a given total energy in two domains of different, but constant potentials.

In the two domains we need different wave vectors, corresponding to the electrons' classical momenta at constant total energy, but we need the same frequency. The latter applies also to all other wave phenomena and is a necessary condition if we are to combine the wave functions in the two domains in such a way that they remain continuous in time. Otherwise the combined wave function could not be a solution to a differential equation in space and time.

Apparently the demand for identical frequencies with different wave vectors is not satisfied by the expression (1.2.12) for the phase, which followed from (1.2.11) with $C = 0$. However, we may manage by suitable adjustment of the arbitrary constant C in the frequencies for the two domains. Within these the wave vectors are known from $\mathbf{k} = \mathbf{p}/\hbar$ and $\mathbf{k}' = \mathbf{p}'/\hbar$. If we let the potential energies be Φ_0 and Φ'_0, and the total energy be E, we find from (1.2.11) for the frequencies of the two domains

$$\omega = \frac{1}{\hbar}(E-\Phi_0) + C,$$

$$\omega' = \frac{1}{\hbar}(E-\Phi'_0)+C'.$$

The desired result is then obtained by putting $C = \Phi_0/\hbar$, $C' = \Phi'_0/\hbar$, whence the frequency in both domains is given by the simple expression

$$\omega = E/\hbar. \tag{1.2.17}$$

The previously mentioned arbitrariness in the absolute magnitude of the frequency now corresponds perfectly to the freedom of choice of the zero point for potential energy.

The inclusion of the extra term, $C = \Phi_0/\hbar$, in (1.2.11) causes the presence of an extra term, $-\Phi_0/\hbar\, t$, in the phase (1.2.12). The wave function $\exp(i\varphi)$ is no longer a solution of (1.2.13), but satisfies instead the differential equation

$$-\frac{\hbar^2}{2m}\Delta\psi+\Phi_0\psi = i\hbar\frac{\partial\psi}{\partial t}. \tag{1.2.18}$$

Correspondingly the wave function inside the crystal satisfies

$$-\frac{\hbar^2}{2m}\Delta\psi+\Phi'_0\psi = i\hbar\frac{\partial\psi}{\partial t}. \tag{1.2.19}$$

One may now be tempted to combine (1.2.18) and (1.2.19) in a single differential equation

$$-\frac{\hbar^2}{2m}\,\Delta\psi+\Phi(\mathbf{r})\psi = i\hbar\,\frac{\partial\psi}{\partial t}\,, \tag{1.2.20}$$

where $\Phi(\mathbf{r})$ represents the potential energy at the point determined by the vector $\mathbf{r}$. Obviously this differential equation comprises (1.2.18) and (1.2.19), since we assumed $\Phi(\mathbf{r}) = \Phi_0$ and $\Phi(\mathbf{r}) = \Phi_0{}'$ for the two domains. Similarly (1.2.20) may be seen to lead to (1.2.13) when, for a free particle, we make the usual assumption $\Phi(\mathbf{r}) = 0$ for all $\mathbf{r}$.

We shall henceforth assume that a particle with a mass m, in any conservative force field with potential energy $\Phi(\mathbf{r})$ may be associated with a wave function $\psi(\mathbf{r},t)$ which is a solution to (1.2.20). This differential equation is called the Schrödinger equation for particles with mass m in the field $\Phi(\mathbf{r})$. However, our arguments for the validity of (1.2.20) are applicable only for the specific case of Φ being piecewise constant in space. In the more general case the Schrödinger equation is, of course, not satisfied by wave functions representing plane harmonic wave trains or wave packets built from them. Whether or not the solutions provide correct descriptions of natural phenomena in such cases can be decided only by experimental tests of the deduced results.

In the following we base wave mechanics on the Schrödinger equation (1.2.20). Subsequently we shall show that it leads to the Newtonian laws of motion for particles in macroscopic force fields.

Problems

1.2.1 Write down the expression corresponding to (1.2.7) for development in time of a one-dimensional wave packet. Use the expression (1.2.11) for the frequency, and show that the shape of the wave packet may be considered unchanged only for a duration which is small compared to $ma^2/\hbar$, where a represents the order of magnitude of the extent of the wave packet. Find this critical time for the case of $m = 1$ g, $a = 10^{-4}$ cm.

1.2.2 Show that by superposition of one dimensional plane waves, A exp $(ik'x)$, with a weight function

$$g(k') = \exp\left(-\frac{(k'-k)^2}{2\delta^2}\right),$$

we obtain a wave packet of the form

$$\psi(x,0) = A\sqrt{2\pi}\,\delta \exp(ikx) \exp\left(-\frac{x^2}{2a^2}\right),$$

where $a = 1/\delta$.

1.2.3 Verify that, for a constant potential $\Phi(\mathbf{r}) = \Phi_0$, (1.2.20) is satisfied by any wave function

$$\psi = A \exp[(i/\hbar)\mathbf{p}\cdot\mathbf{r} - (i/\hbar)Et],$$

where $\mathbf{p}$ and E represent corresponding values of the classical momentum and total energy for particles of mass m.

1.3 INTERPRETATION OF THE WAVE FUNCTION

Before we can begin deriving quantitative results from the Schrödinger equation we need more well-defined ideas on the translation of the properties of the wave function into specifications as to the physical state of the particles. In the course of the discussion on electron diffraction experiments it was mentioned that the wave function $\psi(\mathbf{r},t)$ should give a certain measure of the occurrence of particles at $\mathbf{r}$ at the time t. A simple, non-negative quantity which may be formed from the complex wave function is the absolute square $|\psi|^2 = \psi^*(\mathbf{r},t)\,\psi(\mathbf{r},t)$, and it may therefore appear to be reasonable to let $|\psi(\mathbf{r},t)|^2$ tentatively represent the density of particles at time t at the point $\mathbf{r}$. According to this assumption, then, it is the variation in space of $|\psi|^2$ which determines the distribution of darkening on the photographic plate in a diffraction experiment.

However, the characteristics of the interference picture turn out to be independent of the intensity used. Even if the intensity is so low that only one electron at a time passes the crystal, the same interference picture will be built up over a long period of time as by brief irradiation of high intensity. Hence the interference effects are not dependent on the presence of a large number of electrons in the crystal and cannot be ascribed to an interaction between particles in the incident beam. Consequently the wave function must be associated with the single electron, and a reassessment of our interpretation of $|\psi|^2$ as a density is necessary.

If this interpretation were to be maintained in the literal, classical sense, one would have to imagine the mass of the electron dispersed in space, and in a diffraction experiment at low intensity one should see an individual electron gradually exercising an effect dispersed over the darkened area of the photographic plate. In actual fact the individual electrons are registered as well-localized particles, each of which affects only one or a few of the grains in the photographic plate. This is seen even more clearly if the photographic plate is replaced by counters for the registration of the electrons. Thus, it is quite impossible to demonstrate the interference effects with a single electron; they only gradually appear after many passages of individual electrons.

These experimental facts may be met by a statistical interpretation of the wave function. Let us assume that $|\psi(\mathbf{r},t)|^2 d\tau$ is proportional to the probability of finding electrons at the time t in the volume element $d\tau$ at $\mathbf{r}$. The wave description then serves to calculate, for the various units of the domain of registration, the probability of observing the presence of an electron within this region. When shooting off each individual electron we can indicate areas on the photographic plate which it is likely to hit, and others which it has no chance of reaching, and we can also calculate the distribution of darkening resulting from the passage of many electrons. On the other hand, we cannot accurately predict where the individual electron will hit the plate. The statistical interpretation of the wave function forces us to renounce the strict determinism which forms such an integral part of the Newtonian mechanics. In return, we avoid metaphysical questions as to whether electrons 'are' waves or particles, and we achieve a natural and unassuming description of experimental experience such as the above.

Before we finally accept the interpretation of $|\psi|^2$ as a probability density $P(\mathbf{r},t)$ in space, we must ascertain that this does not bring us into conflict with assumptions already made, and see whether it causes new demands on the formalism. We see immediately that the latter is the case; because the probability of finding the particle somewhere in space must be unity, we must demand that

$$\int P(\mathbf{r},t)d\tau = \int |\psi(\mathbf{r},t)|^2 d\tau = 1, \tag{1.3.1}$$

where the integration is extended throughout space.

In order to allow the above interpretation of $|\psi|^2$, the wave function must satisfy (1.3.1); it is then said to be normalized. A wave function which at a certain time is a square integrable function of the space coordinates may be normalized by multiplication by a suitable constant. As a one dimensional example we may quote the function considered in the previous section,

$$\psi(x) = A \frac{2 \sin (x\delta k)}{x} \exp (ikx), \tag{1.3.2}$$

which may be normalized by a suitable choice of A. The quantity $|\psi|^2$ must, in this case, be interpreted as the probability density on the x-axis and the normalizing integral (1.3.1) is defined correspondingly (cf. problem 1.3.1).

However, we have also encountered wave functions (1.2.14) which do not tend to zero at large distances and which therefore cannot be normalized. Later we shall see that wave functions of this type (occasionally termed improper wave functions) are convenient for the treatment of many problems. This is true particularly of the problem of scattering where one is interested in describing particles coming from a distant part of space, which after interaction disappear again into the distance. For such wave functions, $|\psi|^2$ may be regarded as a relative measure of the probability of finding the particle in regions of equal size. The use of improper wave functions is, however, more of a mathematical convenience than a physical necessity. An infinitely extended wave train such as (1.2.14) may be regarded as a limiting case of a very large wave packet. The infinite wave train may be replaced by a wave function which in all areas of physical interest is virtually identical with the improper wave function, but which at very great distances disappears sufficiently rapidly for the integral (1.3.1) to exist. This type of wave function arises, e.g. from the one dimensional wave packet (1.3.2) by choosing a sufficiently low value of δk.

Temporarily ignoring the use of improper wave functions, we now assume that the wave function disappears at infinity, enabling (1.3.1) to be fulfilled for a certain time. However, the change in time of the wave function is determined by the Schrödinger equation (1.2.20), and we must therefore make sure that (1.3.1) can remain fulfilled. This again requires that

$$\frac{d}{dt} \int |\psi|^2 \, d\tau = 0, \tag{1.3.3}$$

so that particles are neither created nor destroyed.

Considering first a finite domain Ω in space, we find

$$\frac{\mathrm{d}}{\mathrm{d}t}\int_{\Omega} |\psi|^2 \,\mathrm{d}\tau = \int_{\Omega}\left(\frac{\partial \psi^*}{\partial t}\psi + \psi^* \frac{\partial \psi}{\partial t}\right)\mathrm{d}\tau. \qquad (1.3.4)$$

From the Schrödinger equation (1.2.20) we get

$$\frac{\partial \psi}{\partial t} = \frac{1}{i\hbar}\left\{-\frac{\hbar^2}{2m}\Delta\psi + \Phi(\mathbf{r})\psi\right\},$$

$$\frac{\partial \psi^*}{\partial t} = -\frac{1}{i\hbar}\left\{-\frac{\hbar^2}{2m}\Delta\psi^* + \Phi(\mathbf{r})\psi^*\right\},$$

which, by substitution in (1.3.4), gives

$$\frac{\mathrm{d}}{\mathrm{d}t}\int_{\Omega} |\psi|^2 \,\mathrm{d}\tau = \frac{\hbar}{2im}\int_{\Omega} [(\Delta\psi)^*\psi - \psi^*\Delta\psi \,\mathrm{d}\tau]. \qquad (1.3.5)$$

We now make use of Green's theorem, according to which the following equation is valid for two functions, $u(r)$ and $w(r)$:

$$\int_{\Omega} (u\Delta w - w\Delta u)\,\mathrm{d}\tau = \int_{F} (u \operatorname{grad} w - w \operatorname{grad} u)\cdot \mathbf{df}, \qquad (1.3.6)$$

where the integral on the right-hand side is extended over the boundary surface, F, of the domain Ω, and the vector **df** is defined as an outward normal of F.*

Introduction of (1.3.6) in (1.3.5) with $u = \psi$ and $w = \psi^*$ gives

$$\frac{\mathrm{d}}{\mathrm{d}t}\int_{\Omega} |\psi|^2 \,\mathrm{d}\tau = \frac{\hbar}{2im}\int_{F} (\psi \operatorname{grad} \psi^* - \psi^* \operatorname{grad} \psi)\cdot \mathbf{df}. \qquad (1.3.7)$$

When the surface F expands to infinity, the right-hand side of (1.3.7) vanishes, due to our assumption of the normalizability of the wave function, and hence the relation (1.3.3) follows.

We have now made it plausible that $|\psi|^2$ can be identified with a probability density, $P(\mathbf{r},t)$, in space. On this basis it is possible to

---

* The relation (1.3.6) is easily demonstrated by using

$$\operatorname{div}(u \operatorname{grad} w) = u\Delta w + \operatorname{grad} u \cdot \operatorname{grad} w,$$
$$\operatorname{div}(w \operatorname{grad} u) = w\Delta u + \operatorname{grad} u \cdot \operatorname{grad} w,$$

and by then using the Gauss theorem on the vector field $u \operatorname{grad} w - w \operatorname{grad} u$.

find mean values, or expectation values, for all quantities depending on the position of the particle in space. Thus, the mean value for the coordinate x is given by

$$\langle x\rangle = \int P(\mathbf{r},t)x\,\mathrm{d}\tau = \int |\psi(\mathbf{r},t)|^2 x\,\mathrm{d}\tau. \tag{1.3.8}$$

The expectation value $\langle x\rangle$ represents the mean value of a large number of independent measurements of the x coordinate for a particle whose state is described by the wave function $\psi(\mathbf{r},t)$. It is to be noted that the definition (1.3.8) is mathematically equivalent to the classical mechanical definition of the abscissa of the centre of gravity for a mass distribution proportional to $P(\mathbf{r},t)$. Correspondingly, one speaks of $P(\mathbf{r},t)$ as a probability distribution in space at the time t.

By analogy with (1.3.8) we may evaluate the mean value $\langle x^2\rangle$ or, for an arbitrary function of the coordinates,

$$\langle f(\mathbf{r})\rangle = \int |\psi|^2 f(\mathbf{r})\,\mathrm{d}\tau. \tag{1.3.9}$$

In this way one can, for example, define the mean value of the potential energy, $\langle\Phi(\mathbf{r})\rangle$.

Expectation values can be particularly useful for the characterization of wave packets. It is evident that for a state corresponding to a wave packet within narrow bounds around a point $\mathbf{r}_0$, the following relation holds approximately for the mean value of the coordinates:

$$\langle\mathbf{r}\rangle \approx \mathbf{r}_0.$$

Thus, the mean value of the position vector is a well-defined quantity, which may conveniently indicate the location of the wave packet as a whole. If the wave packet is exceedingly narrow, one may accurately determine not only $\langle x\rangle \approx x_0$ but also $\langle x^2\rangle \approx {x_0}^2$, and on the whole $\langle f(x)\rangle \approx f(x_0)$.

While the mean value of the deviation of x from $\langle x\rangle$ (i.e. the quantity $\langle x-\langle x\rangle\rangle$) is zero, the mean value of the square deviation $\langle(x-\langle x\rangle)^2\rangle$ is a measure of the width of the wave packet in the direction of the x-axis. The square root of this quantity

$$\begin{aligned}\sigma_x &= \{\langle(x-\langle x\rangle)^2\rangle\}^{\frac{1}{2}} \\ &= \{\langle x^2\rangle - \langle x\rangle^2\}^{\frac{1}{2}}\end{aligned} \tag{1.3.10}$$

is the root mean square deviation or the fluctuation in x.

Problems

1.3.1 Determine the normalization for the resulting one-dimensional wave function in problem 1.2.2 and for the wave function $\psi(x) = \exp(ikx) \sin(x\delta)/x$. Outline the corresponding probability distributions on the x-axis.

1.3.2 Find the mean values $\langle x \rangle$, $\langle x^2 \rangle$, and the fluctuation σ_x for the states described by the wave functions mentioned in the above problem.

1.4 MOMENTUM DISTRIBUTION

We have seen in the previous section that the form of the wave function at a certain given time provides information, if only of a statistical nature, on the localization of the particle in space. We shall now see how one can introduce a connection between the wave function and the immediate state of motion of the particle.

We have already done this in a specific case by assuming that a free particle of a certain momentum $\mathbf{p}$ should be associated with a plane, harmonic wave with wave vector $\mathbf{k} = \mathbf{p}/\hbar$. The corresponding (improper) wave function is at a chosen time, t, of the form

$$\psi(\mathbf{r}) = A \exp[(\mathrm{i}/\hbar)\, \mathbf{p} \cdot \mathbf{r}], \tag{1.4.1}$$

where A is a complex constant.

A wave function formed by the superposition of harmonic waves with various wave vectors cannot, according to the above rules, be associated with any definite momentum. However, it is possible for such a state to introduce a statistical interpretation, according to which the particle has a certain probability of appearing with each of the momenta included in the superposed wave function. The relative probabilities are determined by the weight contributed to the wave function by the harmonic waves.

As a simple example we may consider the one dimensional wave function

$$\psi(x) = A \exp(\mathrm{i}kx)\, \frac{2 \sin(x\delta k)}{x}, \tag{1.4.2}$$

which, as previously seen, is formed by superposition of harmonic waves

$$\psi(x) = A \int_{k-\delta k}^{k+\delta k} \exp(\mathrm{i}k'x)\, \mathrm{d}k'. \tag{1.4.3}$$

We have here summed the harmonic waves with equal amplitudes and with wave numbers from an interval $k-\delta k$ to $k+\delta k$, corresponding to momenta in the interval

$$\hbar(k-\delta k) < p_x < \hbar(k+\delta k). \tag{1.4.4}$$

In a state with the wave function (1.4.2) we must imagine that there is equal probability of finding the particle with any of the momenta in the interval (1.4.4).

By generalizing this rule it is possible to associate with every normalizable wave function a probability distribution for momentum. According to the Fourier integral theorem, any such function may at a fixed time be expanded into harmonic functions of the type (1.4.1). For a function, $\psi(x)$, of one variable the Fourier expansion is given by the reciprocal relations

$$\psi(x) = \int_{-\infty}^{\infty} g(k) \exp(ikx)\, dk, \tag{1.4.5}$$

$$g(k) = \frac{1}{2\pi}\int_{-\infty}^{\infty} \psi(x) \exp(-ikx)\, dx.$$

A wave function in space, $\psi(\mathbf{r},t)$, at a fixed time, t, is a function of three variables; its Fourier expansion is determined by

$$\psi(\mathbf{r},t) = \int g(\mathbf{k},t) \exp(i\mathbf{k}\cdot\mathbf{r})\, d^3k \tag{1.4.6}$$

with

$$g(\mathbf{k},t) = \frac{1}{(2\pi)^3}\int \psi(\mathbf{r},t) \exp(-i\mathbf{k}\cdot\mathbf{r})\, d^3r, \tag{1.4.7}$$

where the volume integrals in (1.4.6) and (1.4.7) are extended throughout the coordinate space and the **k**-space.

It is convenient to introduce as a new variable in (1.4.6) and (1.4.7) the momentum, $\mathbf{p} = \hbar\mathbf{k}$, corresponding to a wave vector **k**. If, in addition, we introduce the function $f(\mathbf{p},t) = (2\pi)^3 h^{-3/2}$ $g(\mathbf{p}/\hbar, t)$, we may write (1.4.6) and (1.4.7) in a symmetrical form

$$\psi(\mathbf{r},t) = \frac{1}{h^{3/2}}\int f(\mathbf{p},t) \exp[(i/\hbar)\,\mathbf{p}\cdot\mathbf{r}]\, d^3p, \tag{1.4.8}$$

$$f(\mathbf{p},t) = \frac{1}{h^{3/2}}\int \psi(\mathbf{r},t) \exp[(-i/\hbar)\,\mathbf{p}\cdot\mathbf{r}]\, d^3r, \tag{1.4.9}$$

where d^3p is the volume element in momentum space.

The representation (1.4.8) of the wave function $\psi(\mathbf{r},t)$ includes the harmonic waves corresponding to the various momenta $\mathbf{p}$, with complex amplitudes $f(\mathbf{p},t)$, so that the harmonic waves have different weights.

We will assume that the quantity

$$P(\mathbf{p},t) = |Af(\mathbf{p},t)|^2\,, \qquad (1.4.10)$$

where A is independent of $\mathbf{p}$, represents the distribution of momentum (the probability density in momentum space) for a state described by the wave function $\psi(\mathbf{r},t)$ in (1.4.8). Hence the quantity $|f(\mathbf{p},t)|^2\,\mathrm{d}p_x\,\mathrm{d}p_y\,\mathrm{d}p_z$ must be proportional to the probability of finding the particle, at the time t, with momentum components in intervals of the magnitudes $\mathrm{d}p_x$, $\mathrm{d}p_y$ and $\mathrm{d}p_z$ around $\mathbf{p}$. The function $A\,f(\mathbf{p},t)$ is referred to as a probability amplitude.

The interpretation of (1.4.10) as a probability distribution for momentum is possible if we can choose A in such a way that the following is valid:

$$|A|^2 \int |f(\mathbf{p},t)|^2\,\mathrm{d}^3p = 1\,. \qquad (1.4.11)$$

However, as we shall see presently, the fulfilment of (1.4.11) makes no new demands on our formalism. We find that the integral in (1.4.11), using (1.4.9), may be written

$$\int f^* f\,\mathrm{d}^3p = \int f^*\,\frac{1}{h^{3/2}} \int \psi \exp\left[(-\mathrm{i}/\hbar)\,\mathbf{p}\cdot\mathbf{r}\right]\mathrm{d}^3r\,\mathrm{d}^3p. \qquad (1.4.12)$$

If we now assume that we are justified in reversing the order of integration in (1.4.12), and using that according to (1.4.8)

$$\frac{1}{h^{3/2}} \int f^* \exp\left[(-\mathrm{i}/\hbar)\,\mathbf{p}\cdot\mathbf{r}\right]\mathrm{d}^3p = \psi^*,$$

we find from (1.4.12)

$$\int f^* f\,\mathrm{d}^3p = \int \psi^*\psi\,\mathrm{d}^3r, \qquad (1.4.13)$$

from which is seen that (1.4.11) is fulfilled with $A = 1$, when the wave function $\psi(\mathbf{r},t)$ is normalized according to (1.3.1).

The validity of (1.4.13) makes it reasonable to accept the interpretation of $|f(\mathbf{p},t)|^2$ as a probability distribution for momentum, which may be determined from the wave function by the relations (1.4.9) and (1.4.10). In the above treatment we chose our starting point in the description of a free particle, for which the association of momentum with a specific wave function was assumed to be known. We have made no further use of the fact that the particle was free, and we will assume that the connexion introduced above between wave function and momentum distribution may be maintained quite generally. The wave function as well as the momentum distribution will, of course, change in time in a manner depending on the field acting on the particles; but when, at a given time, the wave function varies in a certain way in space, this always corresponds to the same distribution of momentum. This assumption is contained in the more general one, viz., that the wave function at a given time contains the maximum attainable knowledge of the mechanical state of the particle at the chosen time.

It is evident that on the basis of the momentum distribution, $P(\mathbf{p},t)$, it is possible to find mean values for functions of the momentum, in a manner similar to the one used in the preceding section for functions of the space coordinates. Thus, the mean value of the x-component of the momentum at the time t is

$$\langle p_x \rangle = \int p_x \, |f(\mathbf{p},t)|^2 \, d^3p,$$

and the mean value for the kinetic energy is

$$\langle T \rangle = \frac{1}{2m} \int (p_x{}^2 + p_y{}^2 + p_z{}^2) \, |f|^2 \, d^3p.$$

The fluctuation in a momentum component is defined as

$$\sigma_{p_x} = \{\langle p_x{}^2 \rangle - \langle p_x \rangle^2\}^{\frac{1}{2}}.$$

Problems

1.4.1 Find the distribution in momentum p_x corresponding to the one-dimensional wave function

$$\psi(x) = \sqrt{\frac{a}{\pi}} \exp(ikx) \cdot \frac{\sin(x/a)}{x},$$

and calculate for this state the mean values $\langle p_x \rangle$ and $\langle p_x{}^2 \rangle$, as well as the fluctuations σ_{p_x} and σ_x.

1.4.2 Find the distribution in momentum p_x corresponding to the one-dimensional wave function

$$\psi(x) = \sqrt{\frac{2}{\pi}} \, a^{3/2} \frac{\exp(ikx)}{(a^2 + x^2)}.$$

1.4.3 Give a three-dimensional wave function for a state in which the x-component of the momentum has the definite value a, while the y- and z-components are distributed with constant probability density in intervals of magnitude δ and with mean values $\langle p_y \rangle = b$, $\langle p_z \rangle = c$.

1.5 THE UNCERTAINTY RELATIONS

Before we proceed to develop the quantum mechanical formalism and to examine its quantitative results in specific cases, we will discuss some conceptional consequences of the rules of interpretation which we introduced in the previous sections. We shall see that if one accepts the wave mechanics with the interpretation rules mentioned, then one is forced to give up certain concepts and assumptions which in the classical mechanics are considered quite obvious and fundamental.

In classical mechanics, the state of a particle at a certain time is established by its position vector and its velocity or momentum, and it is assumed that it is possible, in principle, to measure these quantities with unlimited accuracy. If the force field is known, then one can determine the state of the particle (i.e. its position and velocity) at any later time by solving the Newton equation with the given initial conditions.

In wave mechanics, the state of the particle at a given time ($t = 0$) is described by infinitely many numbers, i.e., by a wave function $\psi_0(\mathbf{r})$. Through the rules of interpretation established above, this function determines both our knowledge of the position of the particle and its momentum at $t = 0$. Otherwise the problem is analogous to the classical one, inasmuch as we may determine the state of the particle (i.e. its wave function) at any later time by solving the Schrödinger equation with the given $\psi_0(\mathbf{r})$ as the initial condition.

The wave function $\psi_0(\mathbf{r})$ depends of course on the outcome of measurements already made. Thus, if we have determined the position of the particle, we must use a ψ_0 which is a wave packet, and the latter must be chosen sufficiently narrow for $|\psi_0|^2$ to be different from zero only in a small region of space within which we know with certainty that the particle is located, when we take the experimental uncertainty into consideration.

However, we saw in 1.2 that in order to build up this type of wave packet it is necessary to superpose harmonic waves with wave numbers which are spread over intervals of a width inversely proportional to the width of the wave packet. The corresponding 'haziness' in the momentum appeared in 1.2 as a vague concept; but according to the rules introduced later in 1.4 we must actually regard the expansion of the wave function into Fourier components as representing a decomposition into a number of momentum states, which may all, with a certain probability, be registered. Hence a narrowing of the wave packet will automatically cause increasing uncertainty in the momentum which one may associate with the particle, until, for an infinitely narrow wave packet, the momentum becomes completely indefinite. Inversely, an approximation to a state with a definite momentum will cause the position of the particle to become increasingly indefinite. In the limit, when the particle has a perfectly well-defined momentum, the wave function is of the type (1.4.1) for which $|\psi|^2$ is constant throughout space.

The quantitative connexion between the uncertainty in a space coordinate and the lowest uncertainty in the corresponding momentum coordinate will obviously depend upon the more detailed definition of these uncertainties. In 1.2 we assumed, with regard to order of magnitude, a connexion of the type $\delta x \gtrsim 1/\delta k_x (= \hbar/\delta p_x)$. One frequently uses the somewhat sharper inequalities

$$\delta x \delta p_x \geqslant h\,, \quad \delta y \delta p_y \geqslant h\,, \quad \delta z \delta p_z \geqslant h, \tag{1.5.1}$$

which are the famous Heisenberg uncertainty relations for position and momentum. According to these, a particle cannot, in any state, be simultaneously ascribed completely well-defined values for a space coordinate and the corresponding momentum coordinate. On the other hand, nothing prevents, for example, both x and p_y being simultaneously fixed with unlimited accuracy.

The easiest way of illustrating this situation is by considering a wave packet, as for instance represented by the wave function

$$\psi(\mathbf{r}) = \frac{\exp(i\mathbf{k}\cdot\mathbf{r})}{\pi^{3/2}\sqrt{\text{abc}}}\;\frac{\sin(x\text{a})}{x}\;\frac{\sin(y\text{b})}{y}\;\frac{\sin(z\text{c})}{z}, \tag{1.5.2}$$

describing a state with $\langle \mathbf{r} \rangle = (0,0,0)$ and $\langle \mathbf{p} \rangle = \hbar\mathbf{k}$.

The probability distributions for coordinate and momenta corresponding to (1.5.2) are given respectively by

$$|\psi|^2 = \frac{1}{\pi^3 \mathrm{abc}} \frac{\sin^2(x\mathrm{a})}{x^2} \frac{\sin^2(y\mathrm{b})}{y^2} \frac{\sin^2(z\mathrm{c})}{z^2} \tag{1.5.3}$$

and

$$|f(\mathbf{p})|^2 = \begin{cases} \dfrac{1}{(2\hbar)^3 \mathrm{abc}}, & \text{when } \mathbf{p} \text{ is inside the area } \Omega \\ 0 \quad, & \text{when } \mathbf{p} \text{ is outside the area } \Omega, \end{cases} \tag{1.5.4}$$

where Ω is characterized by

$$\begin{aligned} (k_x - \mathrm{a})\hbar &< p_x < (k_x + \mathrm{a})\hbar \\ (k_y - \mathrm{b})\hbar &< p_y < (k_y + \mathrm{b})\hbar \\ (k_z - \mathrm{c})\hbar &< p_z < (k_z + \mathrm{c})\hbar. \end{aligned} \tag{1.5.5}$$

The total uncertainties in the momentum components are, according to (1.5.4) and (1.5.5),

$$\delta p_x = 2\mathrm{a}\hbar, \quad \delta p_y = 2\mathrm{b}\hbar, \quad \delta p_z = 2\mathrm{c}\hbar.$$

For the uncertainties in coordinates no correspondingly sharp measure exists in this case; but it is possible, for example, to measure the uncertainties by the widths of the intervals around the centre of the wave packet, within which $|\psi|^2$ is different from zero. One then finds

$$\delta x = \frac{2\pi}{\mathrm{a}}, \quad \delta y = \frac{2\pi}{\mathrm{b}}, \quad \delta z = \frac{2\pi}{\mathrm{c}},$$

so that in this case we have

$$\delta x \delta p_x = \delta y \delta p_y = \delta z \delta p_z = 2h$$

in agreement with (1.5.1).

Mathematically, the uncertainty relations follow from the fact that the wave function, ψ, and the probability amplitude of the momentum, f, are each other's Fourier transform. By an exact deduction of the uncertainty relations the uncertainties may conveniently be represented by the previously introduced fluctuations for coordinates and momentum. It may be shown that for these the following inequalities are valid:

$$\sigma_x \sigma_{p_x} \geqslant \frac{\hbar}{2}, \quad \sigma_y \sigma_{p_y} \geqslant \frac{\hbar}{2}, \quad \sigma_z \sigma_{p_z} \geqslant \frac{\hbar}{2}, \tag{1.5.6}$$

but we will not here give details of the proof which may be found in any major textbook on quantum mechanics. The difference between the relations (1.5.1) and (1.5.6) which is expressed by a factor of $4\pi \approx 12$ in the value of the minimal uncertainty products, is merely a reflexion of the fact that the uncertainties defined by the fluctuations are smaller than the 'total' uncertainty intervals included in (1.5.1) [see problem 1.4.1].

We will now consider in greater detail the physical consequences of the uncertainty relations. If one accepts the quantum mechanics as a correct theory, then one must conclude that no measuring method allows a determination of the momentum and the position of a particle simultaneously with a greater accuracy than that allowed by the uncertainty relations. If more accurate measurements were possible, they would establish a state which could not be described by any wave function. The limitation in the accuracy of the measurements must be assumed to be of a completely fundamental nature, and thus not to be overcome by any conceivable refinements of the experimental technique.

This type of assumption is altogether unfamiliar in classical mechanics, where the accuracy of the measurements is regarded as a technical matter not subject to any fundamental limitations. This is a basic concept for the classical deterministic philosophy, according to which the future development of all mechanical processes are thought in principle to be predictable with complete certainty, if only it were possible to determine all coordinates and momenta of all particles in the whole 'world' with sufficient accuracy. With the acceptance of the uncertainty relations we are forced to discard completely the concept of particle trajectories of classical physics for systems in which the magnitude of the uncertainty product cannot be neglected. For a hydrogen atom in its ground state the electron has a kinetic energy corresponding to a momentum of a magnitude $\approx 2 \times 10^{-19}$ g cm/sec. In order to justify the concept of a trajectory for the electron, the uncertainties in the momentum components must be small compared to this quantity; but according to the uncertainty relations the space coordinates will then have an uncertainty exceeding $\approx 6\times10^{-27}\times\frac{1}{2}\times10^{19}$ cm $= 3\times10^{-8}$ cm, i.e. a value which exceeds the full extent of the hydrogen atom.

The fundamental limitation in accuracy, for experiments designed to determine simultaneous values of a position coordinate and a momentum coordinate, was the subject of heated discussions among physicists during the years following the advent of the new quantum mechanics. In this connexion a number of imaginary experiments were discussed which particularly clearly demonstrate the physical significance of the uncertainty relations.

As an example we will consider Heisenberg's 'γ-ray microscope', an extremely idealized instrument which is outlined in Fig. 1.5. We imagine a beam of particles of known and well-defined momentum (and hence of indefinite localization) coming in below the microscope. The particles are illuminated by the lamp L in such a way that the scattered light from a particle may be concentrated by the lens to form an image on the screen S. On the basis of the position of the image on the screen, it is possible to determine, for example, the x-coordinate of a particle at a certain time.

However, since the light is in the nature of waves, the image of even an object in the form of a point will be a small spot, i.e.

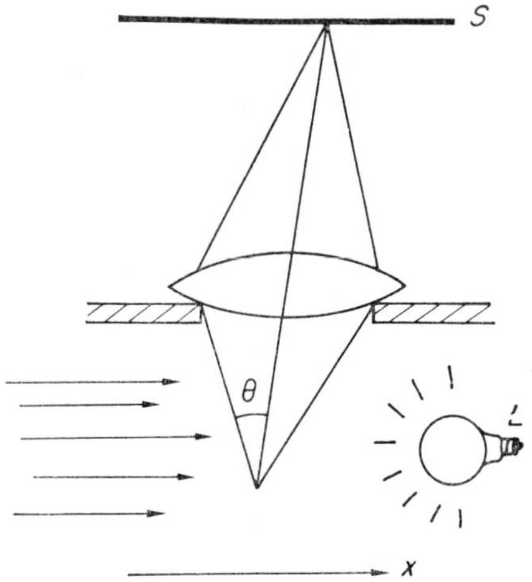

Fig. 1.5

the x-coordinate cannot be determined with an accuracy exceeding

$$\delta x = \frac{\lambda}{2\theta}, \tag{1.5.7}$$

where λ is the wavelength of the light used and θ is half the angle of aperture for the light packet which is being scattered into the microscope from the object. The expression (1.5.7) for the resolution of the microscope is valid for small angles θ; it is derived in optics. Even taking (1.5.7) into account one can still obtain very accurate measurements of coordinates simply by using light of sufficiently short wavelengths, if necessary γ-radiation.

However, it proves to be impossible to make measurements without altering the momentum of the particle by an amount which cannot be determined with an arbitrary accuracy, since the coordinate measurement requires a minimum of one photon to be scattered on the particle and thrown into the microscope. This photon, which has a momentum of magnitude h/λ, transfers a momentum to the particle which may be calculated from the momentum balance if we know the momentum of the photon before and after the collision*. However, after the collision the x-component of the momentum of the photon is uncertain with an amount $\delta p_x \approx 2\theta h/\lambda$, the photon merely being known to have left the particle in some direction within the aperture cone. Hence there is corresponding uncertainty in the momentum which can be ascribed to the particle immediately following the measurement. Consequently we have for this momentum

$$\delta p_x \approx \frac{2h}{\lambda}\,\theta. \tag{1.5.8}$$

The uncertainties (1.5.7) and (1.5.8) resulting from the measurements are seen to be in agreement with the inequalities (1.5.1).

One might perhaps consider reducing the uncertainty (1.5.8) by suspending the microscope in a way sufficiently mobile to measure the momentum transferred by the photon to the microscope, thus determining in which part of the cone of aperture the photon is scattered. However, this type of project would require the momentum

* We imagine the particle to be sufficiently heavy that we may disregard the Compton effect in the scattering.

of the microscope prior to the measurement to have been established with an uncertainty smaller than (1.5.8) and as the uncertainty relations must also be assumed to apply to the microscope as a whole, this would make its position prior to the measurement uncertain with an amount larger than (1.5.7). An attempt to account for the momentum transferred to the microscope at a greater accuracy than (1.5.8) will thus in any case increase the uncertainty in the position coordinate measured for the particle by such an amount that the uncertainty relations are not violated.

Finally, we shall show how the validity of the uncertainty relations prevents logical difficulties in the interpretation of diffraction experiments. This has been particularly emphasized by Niels Bohr, who discussed many imaginary experiments, among which we select the following one.

Electrons with a well-defined momentum $\mathbf{p}$ (corresponding to a plane wave train with wavelength $\lambda = h/p$) are sent towards a plate (see Fig. 1.6) provided with two slits which are narrow compared to the wavelength of the electrons. The slits are parallel and placed a distance a apart. On a screen S, at a distance d from the slits, the electrons will form a distribution of darkening corresponding to interference between two cylindrical waves

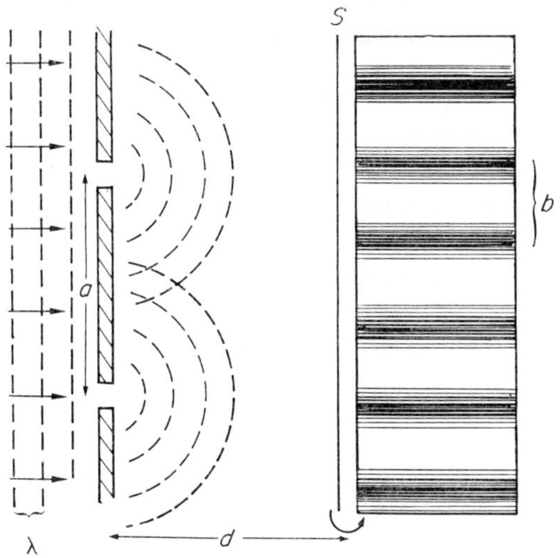

Fig. 1.6

emerging from the slits. The interference pattern will have maximum intensity at points whose distances from the two slits differ by an integral number of wavelengths. The distance between the interference strips is therefore seen to be

$$b \approx d\frac{\lambda}{a} = \frac{d\,h}{p\,a}. \tag{1.5.9}$$

An interference effect is also obtained by the passage of a single electron, since there are regions on the screen which cannot be hit by the electron. This means that the behaviour of the individual electron depends on the presence of both slits, since the interference strips fail to appear in a similar experiment with only one slit where an even distribution of darkening appears on the screen.

Were it now possible in an arrangement with two slits to ensure, without spoiling the interference effects, that the electron had definitely passed through only one of them, e.g. the upper one, then we would be placed in an absurd situation from a logical point of view. This type of difficulty is avoided precisely because of the uncertainty relations, which may be shown to ensure that any experimental arrangement which may serve to identify the slit of passage will prevent observation of the interference effect.

One might, for example, consider identifying the passage slit by measuring the momentum transferred by the electron to the screen at its impact. The vertical momentum component of the electron will thus deviate by an amount $\approx p\,a/d$, according to whether the electron approaches the screen from one or the other of the slits. In order to distinguish between the two possibilities, however, the momentum of the screen must have, in advance, an uncertainty lower than the difference in momentum to be observed. Consequently the momentum uncertainty of the screen must satisfy $\delta p < p\,a/d$, but this causes, according to the uncertainty relations, an uncertainty in the vertical position of the screen of magnitude

$$\delta x \geqslant \frac{h}{\delta p} > \frac{hd}{ap},$$

which in turn, according to (1.5.9), means that in using this arrangement one has *ipso facto* excluded oneself from observing the interference strips.

Problems

1.5.1 Find the value of the fluctuation product $\sigma_x \sigma_{p_x}$ for the two one-dimensional wave functions

$$\text{I} \quad \psi(x) = \left(\frac{2}{\pi}\right)^{\frac{1}{2}} a^{3/2} \exp(ikx) \frac{1}{(a^2+x^2)},$$

$$\text{II} \quad \psi(x) = (2\pi a^2)^{-\frac{1}{4}} \exp(ikx) \exp\left(-\frac{x^2}{4a^2}\right).$$

It should be noted that in the latter case one finds that the sign of equality is valid in (1.5.6). The wave function II is said to represent a minimum wave packet.

1.5.2 A particle with mass m is in such a state that the coordinate fluctuations σ_x, σ_y, and σ_z are all lower than a. Show that the following must be valid for the mean value of the kinetic energy:

$$\langle T \rangle > \frac{3}{8} \frac{\hbar^2}{ma^2}.$$

1.5.3 Use the result from the preceding problem to estimate the lower limit for the kinetic energy of an electron in a hydrogen atom when we assume the extent of the latter to be less than $\approx 10^{-8}$ cm. In a similar way, estimate the minimum kinetic energy required for an electron if it were to be enclosed in an atomic nucleus with radius $\gtrapprox 10^{-12}$ cm.

1.6 OPERATORS FOR MOMENTUM AND ENERGY

We have seen in 1.3 and 1.4 how it is possible for a given state to define mean values of space coordinates and momenta, or functions of them. In the case of the space coordinates the mean values were determined directly from the wave function, as, for example, in (1.3.8). For a momentum component, the mean value was defined indirectly by means of the momentum distribution corresponding to the wave function, which is given by (1.4.9) and (1.4.10). In the following we shall see how one can calculate directly the mean value of momenta also, without in either case having to carry out a Fourier transformation.

The mean value of the momentum component p_x is defined as

$$\langle p_x \rangle = \int p_x \, |f(\mathbf{p})|^2 \, d^3p = \int p_x f^*(\mathbf{p}) f(\mathbf{p}) \, d^3p \qquad (1.6.1)$$

where $f(\mathbf{p})$ is determined from the wave function $\psi(\mathbf{r})$ by (1.4.9). According to (1.4.9) we have

$$f^*(\mathbf{p}) = \frac{1}{h^{3/2}} \int \psi^*(\mathbf{r}) \exp[(i/\hbar)\mathbf{p} \cdot \mathbf{r}] \, d^3r,$$

which by substitution in (1.6.1) and reversal of the order of integration gives

$$\langle p_x \rangle = \int \psi^*(\mathbf{r}) \frac{1}{h^{3/2}} \int f(\mathbf{p}) p_x \exp [(\mathrm{i}/\hbar)\mathbf{p} \cdot \mathbf{r}] \, \mathrm{d}^3 p \, \mathrm{d}^3 r. \qquad (1.6.2)$$

The integral over the momentum space in (1.6.2) differs from the right-hand side of (1.4.8) only by a factor of p_x in the integrand. However, such a factor may easily be introduced into (1.4.8), since the equation

$$p_x \exp [(\mathrm{i}/\hbar)\mathbf{p} \cdot \mathbf{r}] = -\mathrm{i}\hbar \frac{\partial}{\partial x} \exp [(\mathrm{i}/\hbar)\mathbf{p} \cdot \mathbf{r}]$$

is valid.

By partial differentiation of (1.4.8) with regard to x and multiplication by $-\mathrm{i}\hbar$ we then get

$$-\mathrm{i}\hbar \frac{\partial}{\partial x} \psi = \frac{1}{h^{3/2}} \int f(\mathbf{p}) p_x \exp [(\mathrm{i}/\hbar)\mathbf{p} \cdot \mathbf{r}] \, \mathrm{d}^3 p, \qquad (1.6.3)$$

which by substitution in (1.6.2) gives

$$\langle p_x \rangle = \int \psi^*(\mathbf{r}) \left(-\mathrm{i}\hbar \frac{\partial}{\partial x} \right) \psi(\mathbf{r}) \, \mathrm{d}^3 r. \qquad (1.6.4)$$

Thus, we have achieved an explicit expression for $\langle p_x \rangle$ in terms of the wave function. In a perfectly analogous manner we may express the mean value of an arbitrary integral power of p_x. To this end we use the relation

$$p_x{}^n \exp [(\mathrm{i}/\hbar)\mathbf{p} \cdot \mathbf{r}] = \left(-\mathrm{i}\hbar \frac{\partial}{\partial x} \right)^n \exp [(\mathrm{i}/\hbar)\mathbf{p} \cdot \mathbf{r}],$$

and find

$$\langle p_x{}^n \rangle = \int \psi^*(\mathbf{r}) \left(-\mathrm{i}\hbar \frac{\partial}{\partial x} \right)^n \psi(\mathbf{r}) \, \mathrm{d}^3 r. \qquad (1.6.5)$$

We may simplify the expressions (1.6.4) and (1.6.5) by introducing the operator

$$\hat{p}_x = -\mathrm{i}\hbar \frac{\partial}{\partial x}, \qquad (1.6.6)$$

which is termed the operator for the x-component of the momentum. The mean value (1.6.5) may then be written

$$\langle p_x{}^n \rangle = \int \psi^* \, \hat{p}_x{}^n \, \psi \, \mathrm{d}^3 r, \tag{1.6.7}$$

where $\hat{p}_x{}^n$ in the integrand indicates that the operator (1.6.6) must act n times on ψ before the result is multiplied by ψ^*.

In the above we have considered only mean values with reference to p_x. It is trivial that one may in an analogous manner express the mean values for powers of p_x and p_z by means of the two momentum operators

$$\hat{p}_y = -\mathrm{i}\hbar \frac{\partial}{\partial y} \quad \text{and} \quad \hat{p}_z = -\mathrm{i}\hbar \frac{\partial}{\partial z} . \tag{1.6.8}$$

Considerations similar to the above may also easily show that mean values of sums and products of the momentum components may be expressed by means of the operators (1.6.6) and (1.6.8), so that we have, for example,

$$\langle p_x{}^n + p_y{}^m \rangle = \int \psi^* (\hat{p}_x{}^n + \hat{p}_y{}^m) \psi \, \mathrm{d}^3 r$$

and

$$\langle p_x{}^n p_y{}^m \rangle = \int \psi^* \hat{p}_x{}^n \, \hat{p}_y{}^m \, \psi \, \mathrm{d}^3 r.$$

From this is seen that for an arbitrary function $F(p_x, p_y, p_z)$ which is in the form of a polynomial in p_x, p_y, and p_z, we may write the mean value of F in the form

$$\langle F(p_x,p_y,p_z) \rangle = \int \psi^* \, F(\hat{p}_x,\hat{p}_y,\hat{p}_z) \, \psi \, \mathrm{d}\tau. \tag{1.6.9}$$

This result may be formulated in the following 'prescription' for the calculation of the mean value of a function of the momentum components. In the function $F(p_x,p_y,p_z)$ substitute for the momentum components the corresponding operators (1.6.6) and (1.6.8), and let the resulting operator act on ψ in the normalizing integral $\int \psi^* \psi \, \mathrm{d}\tau$. The same prescription may be used when calculating the mean value for a function $G(\mathbf{r})$ of the space coordinates, when $G(\mathbf{r})$ is regarded as the trivial operator whose effect on ψ is merely that ψ is multiplied by $G(\mathbf{r})$. On account of the formal similarity to (1.6.9), the mean value of $G(\mathbf{r})$ is usually written in the form

$$\langle G(\mathbf{r}) \rangle = \int \psi^* G(\mathbf{r}) \, \psi \, \mathrm{d}\tau \, , \tag{1.6.10}$$

although in this case the sequence of the factors in the integrand is of no importance.

We will make the reasonable assumption that for a sum of functions of space coordinates and momenta one may define the mean value by

$$\langle G(\mathbf{r})+F(\mathbf{p})\rangle = \langle G(\mathbf{r})\rangle+\langle F(\mathbf{p})\rangle. \tag{1.6.11}$$

One may then calculate the mean value of the total energy for a particle in a conservative force field with potential energy $\Phi(\mathbf{r})$. For the calculation of the mean value of the kinetic energy, $T = (p_x^2+p_y^2+p_z^2)/2\,m$, we must, according to the above, use the operator

$$(\hat{p}_x^2+\hat{p}_y^2+\hat{p}_z^2)/2\,m$$

$$= \frac{(-i\hbar)^2}{2\,m}\left\{\left(\frac{\partial}{\partial x}\right)^2+\left(\frac{\partial}{\partial y}\right)^2+\left(\frac{\partial}{\partial z}\right)^2\right\} = -\frac{\hbar^2}{2\,m}\,\Delta\,.$$

One finds then for the mean value of the total energy E

$$\langle E\rangle = \langle T\rangle+\langle\Phi\rangle = \int \psi^*\left(-\frac{\hbar^2}{2m}\,\Delta+\Phi(\mathbf{r})\right)\psi\,\mathrm{d}\tau. \tag{1.6.12}$$

The operator

$$\hat{H} = -\frac{\hbar^2}{2m}\,\Delta+\Phi(\mathbf{r}), \tag{1.6.13}$$

which enters in (1.6.12), is referred to as the Hamiltonian operator for the particle in the force field in question. It is constructed from the classical Hamiltonian function for the particle

$$H(\mathbf{p},\mathbf{r}) = \frac{p_x^2+p_y^2+p_z^2}{2m}+\Phi(\mathbf{r})$$

by replacing the momentum components by the corresponding operators.

Introducing the Hamiltonian we are able to write the Schrödinger equation (1.2.20) in the simple form

$$\hat{H}\psi = i\hbar\,\frac{\partial\psi}{\partial t}\,. \tag{1.6.14}$$

We will now illustrate the way in which to apply this formalism by calculating the time derivative of $\langle E\rangle$. By differentiation of (1.6.12) we find

$$\frac{\mathrm{d}}{\mathrm{d}t}\langle E\rangle = \frac{\mathrm{d}}{\mathrm{d}t}\int \psi^*\hat{H}\psi\,\mathrm{d}\tau$$

$$= \int \frac{\partial\psi^*}{\partial t}\hat{H}\psi\,\mathrm{d}\tau + \int \psi^*\hat{H}\frac{\partial\psi}{\partial t}\,\mathrm{d}\tau. \tag{1.6.15}$$

The last term of (1.6.15) may be transformed into

$$\int (\hat{H}\psi^*)\frac{\partial\psi}{\partial t}\,\mathrm{d}\tau,$$

where the Hamiltonian has been shifted to operate on ψ^* instead of on $\partial\psi/\partial t$. The justification of this transformation follows from our assumption that the wave function disappears rapidly at infinite distances. One finds in fact that

$$\int \psi^*\hat{H}\frac{\partial\psi}{\partial t}\,\mathrm{d}\tau - \int (\hat{H}\psi^*)\frac{\partial\psi}{\partial t}\,\mathrm{d}\tau \tag{1.6.16}$$

$$= -\frac{\hbar^2}{2m}\int_\Omega \left(\psi^*\Delta\frac{\partial\psi}{\partial t} - (\Delta\psi^*)\frac{\partial\psi}{\partial t}\right)\mathrm{d}\tau \to 0 \text{ for } \Omega \to \infty,$$

the integral being transformed by means of Green's theorem (1.3.6) into a surface integral over a surface at infinite distances (see p. 21).

Thus, we may transform (1.6.15) into

$$\frac{\mathrm{d}}{\mathrm{d}t}\langle E\rangle = \int\left(\frac{\partial\psi^*}{\partial t}\hat{H}\psi + (\hat{H}\psi^*)\frac{\partial\psi}{\partial t}\right)\mathrm{d}\tau, \tag{1.6.17}$$

which, by the use of the Schrödinger equation (1.6.14), becomes

$$\frac{\mathrm{d}}{\mathrm{d}t}\langle E\rangle = \int\left(\frac{\partial\psi^*}{\partial t}\,\mathrm{i}\hbar\frac{\partial\psi}{\partial t} - \mathrm{i}\hbar\frac{\partial\psi^*}{\partial t}\frac{\partial\psi}{\partial t}\right)\mathrm{d}\tau = 0. \tag{1.6.18}$$

For a particle in a conservative force field we thus find that the mean value of the total energy is constant. This plausible result may partly justify the assumption (1.6.11) and the resulting expression for $\langle E\rangle$.

We saw above that the rôle of the Hamiltonian in the calculation of $\langle E \rangle$ is completely analogous to the rôle of the operator $\hat{p}_x$ in a calculation of $\langle p_x \rangle$. We shall then, in the following, assume as a matter of course that the Hamiltonian (1.6.13) may also serve in the calculation of mean values of powers of E by the relation

$$\langle E^n \rangle = \int \psi^* \hat{H}^n \psi \, d\tau. \tag{1.6.19}$$

Thus, the Hamiltonian is introduced as operator for the total energy in complete analogy to the introduction of the operators (1.6.6) and (1.6.8) for the momentum components.

Equation (1.6.16) represents a special case of a more common property of the operator $\hat{H}$, i.e. $\hat{H}$ is a Hermitian operator. A given operator, $\hat{S}$, is said to be Hermitian if the following equation is fulfilled for arbitrary normalizable wave functions ψ_1 and ψ_2:

$$\int \psi_1^* \hat{S} \psi_2 \, d\tau = \int \psi_2 (\hat{S} \psi_1)^* \, d\tau. \tag{1.6.20}$$

The fact that $\hat{H}$ is a Hermitian operator is shown in complete analogy to (1.6.16), and one may show correspondingly by partial integration that the operators $\hat{p}_x$, $\hat{p}_y$ and $\hat{p}_z$ are also Hermitian.

Problem

1.6.1 Show, by using (1.6.4), that for a state described by the one-dimensional wave function in problem 1.4.2 we have $\langle p_x \rangle = \hbar k$.

2

Stationary States

2.1 DETERMINATION OF STATIONARY STATES

In the preceding chapter we have introduced certain basic features of the quantum mechanical formalism. In the present chapter we investigate some special solutions of the Schrödinger equation which represent so-called stationary states, and which play an important part in the treatment of most quantum mechanical problems.

We start by quoting the Schrödinger equation (1.2.20) for a particle in a potential field $\Phi(\mathbf{r})$:

$$\left[-\frac{\hbar^2}{2m}\Delta+\Phi(\mathbf{r})\right]\psi = \mathrm{i}\hbar\frac{\partial\psi}{\partial t}. \tag{2.1.1}$$

In order to obtain stationary solutions we try to get rid of the time dependence. To this end, we tentatively write ψ as a product of two functions, one of which depends only on the space coordinates, the other only on time. When substituting a function of this kind, i.e.

$$\psi(\mathbf{r},t) = u(\mathbf{r})v(t), \tag{2.1.2}$$

in (2.1.1), it is noted that the differential operations on the left-hand and the right-hand side of (2.1.1) act only on u and v, respectively. After division by $\psi = uv$, we get from (2.1.1)

$$\frac{1}{u}\left[-\frac{\hbar^2}{2m}\Delta + \Phi(\mathbf{r})\right]u = \mathrm{i}\hbar\frac{1}{v}\frac{\mathrm{d}v}{\mathrm{d}t}. \tag{2.1.3}$$

Here we have obtained a separation of the variables, so that the left-hand side of (2.1.3) depends on $\mathbf{r}$ only, while the right-hand side

D

depends solely on t*. The identity between the expressions on either side of the equality sign can then be maintained only if both expressions are equal to a constant, $\mathscr{E}$. Hence we may replace (2.1.3) by

$$i\hbar \frac{dv}{dt} = \mathscr{E}v \tag{2.1.4}$$

and

$$\left[-\frac{\hbar^2}{2m}\Delta + \Phi(\mathbf{r})\right] u = \mathscr{E}u. \tag{2.1.5}$$

Of these two differential equations (2.1.4) may immediately be integrated to give

$$v(t) = A \exp\left[-(i/\hbar)\mathscr{E}t\right], \tag{2.1.6}$$

whereas the solution (2.1.5) depends on the behaviour of the potential $\Phi(\mathbf{r})$ in each individual case.

A solution to (2.1.1) of the type (2.1.2) may thus be written in the form

$$\psi(\mathbf{r},t) = u(\mathbf{r}) \exp\left[-(i/\hbar)\mathscr{E}t\right]. \tag{2.1.7}$$

The space part of the wave function, $u(\mathbf{r})$, is then fixed only subject to an arbitrary constant factor (in agreement with (2.1.5) being homogeneous in u); the modulus of this factor may be established by stipulating that the wave function (2.1.7) must be normalized.

The probability density in space corresponding to (2.1.7) is given by

$$|\psi|^2 = |u(\mathbf{r})|^2,$$

which is constant in time. Hence the mean value for any function of the coordinates is constant. This is true also of the mean value of a function of the momentum components. In fact, since the momentum operators (1.6.6) and (1.6.8) act only on the spatial part of the wave function, we find by using (1.6.9) with ψ given by (2.1.7) that

$$\langle F(p_x,p_y,p_z)\rangle = \int u^* F(\hat{p}_x,\hat{p}_y,\hat{p}_z) u \, d\tau,$$

in which the time does not enter. A particle described by a wave function of the type (2.1.7) is said to be in a stationary state. It is characterized by the fact that the physical state of the system does not change in time.

---

* Of course this separation could not be made if the potential Φ were time dependent.

However, there is another important characteristic of stationary states, which is that the total energy in this type of state has a sharply defined value. This is connected with the fact that the spatial part of the wave function satisfies (2.1.5), which in terms of the Hamilton operator (1.6.13) may be written

$$\hat{H}u = \mathscr{E}u. \tag{2.1.8}$$

Hence a function $u(\mathbf{r})$ which satisfies (2.1.5) has the property that when the Hamilton operator acts on $u(\mathbf{r})$ the result is merely the function $u(\mathbf{r})$ itself, multiplied by a constant $\mathscr{E}$. Such a function is called an eigenfunction for the operator $\hat{H}$, corresponding to the eigenvalue $\mathscr{E}$.

Since $\hat{H}$ acts only on the space part of the wave function, one finds $\hat{H}\psi = \mathscr{E}\psi$ when ψ is of type (2.1.7). The mean value of any power of the total energy in a stationary state is then easily calculated:

$$\langle E^n \rangle = \int \psi^* \hat{H}^n \psi \mathrm{d}\tau = \mathscr{E}^n \int \psi^* \psi \mathrm{d}\tau = \mathscr{E}^n, \tag{2.1.9}$$

where the wave function is assumed to be normalized.

This result shows that total energy in the stationary state has the definite value $\mathscr{E}$. It is then realized from (2.1.9) that for each polynomial in E, $F(E)$, the result $\langle F(E) \rangle = F(\mathscr{E})$ applies, and that, specifically, the fluctuation in energy, σ_E, is zero, since (2.1.9) implies $\langle E^2 \rangle = \langle E \rangle^2$. Such a state in which the total energy may be ascribed a definite value is also called an eigenstate of the energy, because it is described by a wave function which is an eigenfunction of the energy operator.

Since we have seen that the separation constant may be identified with the total energy, E, in the stationary state, we write (2.1.7) and (2.1.5) respectively as

$$\psi(\mathbf{r},t) = u(\mathbf{r}) \exp\left[-(\mathrm{i}/\hbar)Et\right], \tag{2.1.10}$$

and

$$\left[-\frac{\hbar^2}{2m}\Delta + \Phi(\mathbf{r})\right] u = Eu. \tag{2.1.11}$$

The wave function for a stationary state with the total energy E is then of the form (2.1.10), where $u(\mathbf{r})$ must satisfy (2.1.11). Hence such a wave function varies harmonically in time with a frequency $\omega = E/\hbar$, and the function $u(\mathbf{r})$ may be regarded as an amplitude varying in space. For the stationary state, the variation in time of the wave function is not connected to any change in the physical state of the system, which is completely characterized by the function u($\mathbf{r}$). This is often referred to as the wave function for the stationary state, leaving out the trivial time factor $\exp(-\mathrm{i}Et/\hbar)$.

An examination of the stationary states of a particle in a potential $\Phi(\mathbf{r})$ must be based on a solution of the differential equation (2.1.11). This appears under various names, such as the time-independent Schrödinger equation, Schrödinger's amplitude equation, or simply the Schrödinger equation, (2.1.1) being then referred to as the time-dependent Schrödinger equation. Since the potential energy, $\Phi(\mathbf{r})$, and the total energy, E, in (2.1.11) are real quantities, the solutions, $u(\mathbf{r})$, may be chosen as real functions. In a complex solution to (2.1.11), the real and the imaginary part must each individually satisfy the differential equation.

As the first example of the determination of stationary states we consider the case $\Phi = 0$, i.e. the free particle. The solutions to the equations (2.1.10) and (2.1.11) are seen to be, in fact, the previously discussed plane waves (1.2.14), and we thus find for each energy $E > 0$ a solution of the time-independent Schrödinger equation

$$u = \exp(\mathrm{i}\mathbf{k}\cdot\mathbf{r}), \tag{2.1.12}$$

only if

$$\frac{\hbar^2(\mathbf{k})^2}{2m} = E. \tag{2.1.13}$$

For each energy value we thus have a large number of linearly independent wave functions. We note that the probability density $|u|^2$ is constant throughout space, and that the stationary states for a free particle are thus described by improper wave functions.

Instead of (2.1.12) we could have used the real wave functions

$$u = \sin(\mathbf{k}\cdot\mathbf{r})$$

or (2.1.14)

$$u = \cos(\mathbf{k}\cdot\mathbf{r}),$$

where $\mathbf{k}$ satisfies (2.1.13). However, one usually describes a free particle by means of the complex functions, since the wave function (2.1.12) describes a state which, in addition to having a definite energy, also possesses a well-defined momentum.

We prove this by showing that the wave function (2.1.12), besides being an eigenfunction of the energy operator $\hat{H}$, is also an eigenfunction of the momentum operators $\hat{p}_x$, $\hat{p}_y$, and $\hat{p}_z$. By letting the momentum operators (1.6.6)–(1.6.8) act on (2.1.13), we find in fact that

$$\begin{aligned} \hat{p}_x u &= \hbar k_x u = p_x u \\ \hat{p}_y u &= \hbar k_y u = p_x u \\ \hat{p}_z u &= \hbar k_z u = p_z u \end{aligned} \tag{2.1.15}$$

or for short

$$\hat{\mathbf{p}} u = \mathbf{p} u, \text{ where } \mathbf{p} = \hbar \mathbf{k}. \tag{2.1.16}$$

By working out $\langle p_x{}^n \rangle$, $\langle p_y{}^n \rangle$ and $\langle p_z{}^n \rangle$, one shows as above [equation (2.1.9)] that these equations imply that all three components of the momentum of state (2.1.12) have the definite values $\hbar k_x$, $\hbar k_y$ and $\hbar k_z$. The states, described by (2.1.14) have no well-defined values of $\mathbf{p}$, but may be regarded as a superposition of states in which $\mathbf{p} = \hbar \mathbf{k}$ and $\mathbf{p} = -\hbar \mathbf{k}$.

Whereas for $\Phi = 0$ (and $E > 0$) we found that any solution of (2.1.11) was a physically acceptable (although improper) wave function, this is not the case in general.

The solution must, after multiplication by $\exp(-\mathrm{i}Et/\hbar)$, represent a wave function for a physical state, which requires the fulfilment of certain boundary conditions for $u(\mathbf{r})$. If we demand the wave function to be normalizable, this means that $u(\mathbf{r})$ must vanish at infinity; but as one wants also to accept improper wave functions, one may only insist that the wave function must be finite throughout space.

The introduction of such an extra condition for the function $u(\mathbf{r})$ implies, as we shall see, that for a given potential field (2.1.11) does not have acceptable solutions for all the values of E which would have been possible according to classical mechanics. One finds frequently that in certain energy intervals only a finite or infinite set of discrete energy values lead to acceptable solutions. An analysis of the differential equation (2.1.11) in specific cases may therefore serve to determine the possible energy states for the particle in the potential in question.

In order to see how one may be led to discrete energy states, we consider the solution of (2.1.11) for a one-dimensional problem characterized by a potential energy $\Phi(x)$, as outlined in the energy diagram of Fig. 2.1. For a particle in such a potential, the classically possible energy values are limited only by $E>\Phi_{\min}$. We now choose an energy in this range, E, and discuss the possibility of finding acceptable solutions of (2.1.11). In the present one-dimensional case this differential equation becomes

$$\frac{\hbar^2}{2m}\frac{\mathrm{d}^2u}{\mathrm{d}x^2} = -[E-\Phi(x)]u, \tag{2.1.17}$$

which is a linear and homogeneous ordinary differential equation of 2nd order.

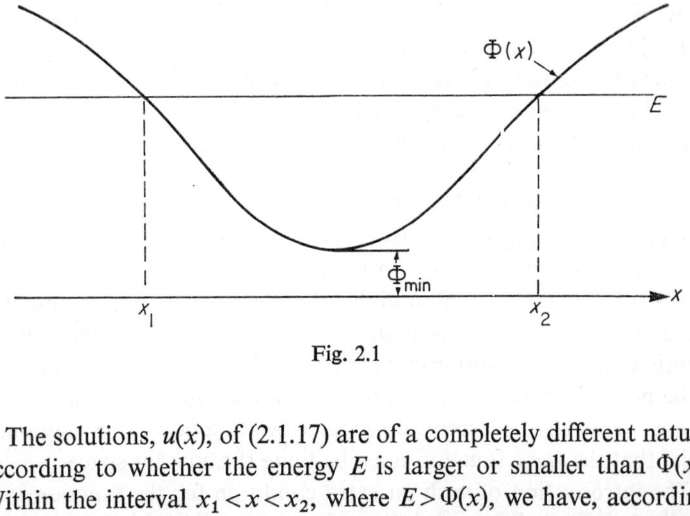

Fig. 2.1

The solutions, $u(x)$, of (2.1.17) are of a completely different nature according to whether the energy E is larger or smaller than $\Phi(x)$. Within the interval $x_1<x<x_2$, where $E>\Phi(x)$, we have, according to (2.1.17), $u''/u<0$, which means that the function $u(x)$ turns its concave side towards the zero line. Hence the function will here have an oscillating character if the interval is sufficiently wide. Within the regions $x<x_1$ and $x>x_2$, where $E<\Phi(x)$, we have $u''/u>0$, and thus the function turns its convex side towards the zero line. As these intervals are unlimited, $|u(x)|$ will here end up by increasing monotonically, unless the function tends asymptotically towards zero.

In order to achieve a $u(x)$ which remains limited for $x\to+\infty$, it is therefore necessary to choose the definite linear combination of the two independent integrals to (2.1.17), which tends asymptotically towards zero for $x\to+\infty$. Thus $u(x)$ is fixed subject to a constant factor. However, this type of solution will not usually satisfy the boundary conditions for $x \to -\infty$, and thus for the E value in question (2.1.11) has no acceptable solution. Only in exceptional cases, for selected values of E, can one find a solution to (2.1.11) which will satisfy the boundary conditions for both $x \to +\infty$ and for $x \to -\infty$. Such a value of E represents a possible energy for a stationary state. In Fig. 2.2 the solutions to (2.1.11) satisfying the boundary conditions for $x \to +\infty$ and for $x \to -\infty$ are outlined for two different values of E. In the upper, typical, case the boundary conditions determine two linearly independent functions. The lower case represents a possible energy state for the particle.

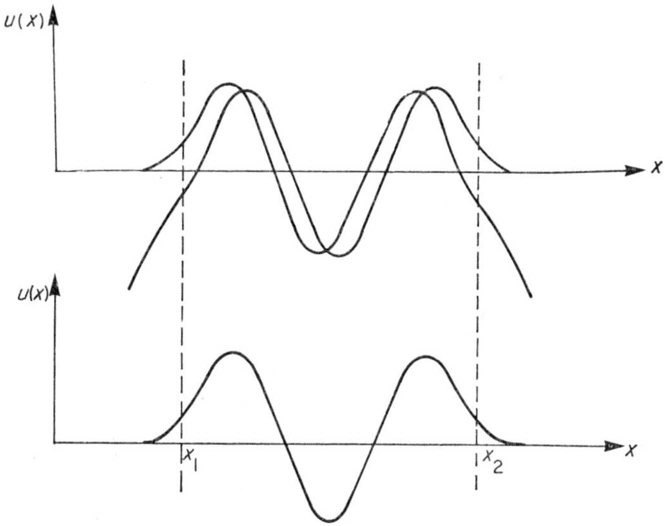

Fig. 2.2

Problem

2.1.1 Show directly, by means of (1.4.9) and (1.4.10), that the momentum distribution for a stationary state does not change with time.

2.2 PARTICLE IN A POTENTIAL BOX

In this and the following section we will examine the stationary states for some potential fields which lead to an especially simple integration of (2.1.11). These are fields where $\Phi(r)$ can be considered to be piecewise constant. We assume that no forces act on the particle except very close to certain surfaces in space. Thus the potential energy is constant, except for a strong variation at these surfaces which can be represented by potential jumps.

Due to the discontinuity of the function, $\Phi(\mathbf{r})$, appearing in the differential equation (2.1.11), one cannot directly find solutions, $u(\mathbf{r})$, in the entire space. One has to find solutions in each of the regions where Φ is constant and afterwards match the solutions in such a way that (2.1.11) is satisfied in the entire space. This puts certain demands of continuity on the wave function on the boundary surfaces, as it is easily shown by considering the problem in one dimension.

With a potential energy $\Phi(x)$ one can write (2.1.11) as

$$\frac{\mathrm{d}^2u}{\mathrm{d}x^2}+\frac{2m}{\hbar^2}[E-\Phi(x)]u = 0 \tag{2.2.1}$$

from which one gets by integration

$$\frac{\mathrm{d}u}{\mathrm{d}x} = -\frac{2m}{\hbar^2}\int^x [E-\Phi(x')]u(x')\mathrm{d}x'.$$

Hence, it is immediately seen that $\mathrm{d}u/\mathrm{d}x$ must be continuous, even though $\Phi(x)$ is discontinuous. It similarly follows that also u is continuous everywhere.

We can utilize this to combine solutions of equation (2.2.1) at a jump between two constant potentials. Put $\Phi(x) = 0$ for $x<0$, $\Phi(x) = \Phi_0>0$ for $x>0$. The differential equation (2.2.1) will then be

$$\text{for } x<0\text{: } u''+k^2u = 0, \text{ with } k = \left(\frac{2mE}{\hbar^2}\right)^{\frac{1}{2}},$$

$$\text{for } x>0\text{: } u''-\beta^2u = 0, \text{ with } \beta = \left(\frac{2m(\Phi_0-E)}{\hbar^2}\right)^{\frac{1}{2}}.$$

We will only consider the case where $E<\Phi_0$ and in the two regions the real solutions are then of the form $\sin(kx+\delta)$ for $x<0$ and $\exp(\pm\beta x)$ for $x>0$. In order that $u(x)$ can remain finite, when $x\to\infty$, we must in region II ($x>0$) discard the increasing exponential function. The solution in II can then be written

$$u_{\mathrm{II}}(x) = A_{\mathrm{II}}\exp(-\beta x). \tag{2.2.2}$$

An arbitrary solution in the region I ($x<0$) can be written

$$u_{\mathrm{I}}(x) = A_{\mathrm{I}} \sin (kx+\delta), \tag{2.2.3}$$

but in order that u_{I} and u_{II} together form a solution, A_{I} and A_{II} must be chosen such that the conditions of continuity for the function and its derivative $u_{\mathrm{I}}(0) = u_{\mathrm{II}}(0)$ and $u_{\mathrm{I}}'(0) = u_{\mathrm{II}}'(0)$, are fulfilled. For the functions (2.2.2) and (2.2.3) the conditions of continuity will be

$$\begin{aligned} A_{\mathrm{I}} \sin \delta &= A_{\mathrm{II}} \\ kA_{\mathrm{I}} \cos \delta &= -\beta A_{\mathrm{II}}, \end{aligned} \tag{2.2.4}$$

and hence

$$\tan \delta = -k/\beta,\ A_{\mathrm{I}} = \left(1+\frac{\beta^2}{k^2}\right)^{\frac{1}{2}} A_{\mathrm{II}}. \tag{2.2.5}$$

In the following we will often be interested in the limiting case, where Φ_0 can be regarded as infinitely large in comparison to E. The behaviour of the wave function at such an infinitely high potential wall can be determined from (2.2.5) by taking the limit $\beta/k \to \infty$, which leads to $\delta = 0$, $A_{\mathrm{II}} = 0$, A_{I} arbitrary. At an infinitely high potential wall it is thus seen that the wave function must vanish.

We will now use this result to discuss the stationary states for a particle which is free to move along the x-axis inside an interval which is bounded by infinitely high potential walls. This extreme case is called a one-dimensional potential box. Let the walls be at $x = -a/2$ and $x = a/2$ and put $\Phi(x) = 0$ for $-a/2<x<a/2$. The differential equation (2.2.1) in the interval $-a/2<x<a/2$ can then be written in the form

$$u''+k^2u = 0, \tag{2.2.6}$$

where

$$k = \left(\frac{2mE}{\hbar^2}\right)^{\frac{1}{2}}. \tag{2.2.7}$$

The solutions to (2.2.6) can be chosen as linear combinations of $\sin kx$ and $\cos kx$. As the wave function must be zero at $x = \pm a/2$, we see that (2.2.6) has acceptable solutions only when $ka/2$ is an integral multiple of $\pi/2$. The possible values of k are then

$$k_n = n\frac{\pi}{a}, \qquad n = 1, 2, 3, \ldots \tag{2.2.8}$$

and the corresponding wave functions in the box have the form

$$u(x) = A_n \begin{cases} \cos k_n x \quad , & \text{for } n \text{ odd} \\ \sin k_n x \quad , & \text{for } n \text{ even.} \end{cases} \tag{2.2.9}$$

Notice that $n = 0$ is omitted in (2.2.8) because the zero solution $u(x) = 0$ cannot describe any state. Also negative integer values of n have been neglected since they would lead to the same states (2.2.9).

From (2.2.8) and (2.2.7) one finds the following possible energy values:

$$E_n = \frac{\hbar^2\pi^2}{2ma^2}\, n^2 = \frac{h^2}{8ma^2}\, n^2, \qquad n = 1, 2, 3 \ldots \tag{2.2.10}$$

which represent an infinite set of discrete energy levels. This is said to constitute the energy spectrum of the system. The integer n, which numbers the levels and the wave functions, is called the quantum number.

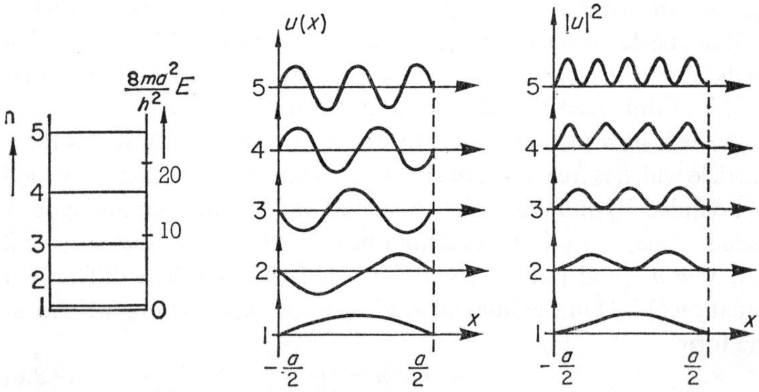

Fig. 2.3

For the states with the lowest quantum numbers, Fig. 2.3 shows the energy levels plotted in an energy diagram (level diagram), together with the corresponding wave functions and their absolute squares, which represent the probability density on the x-axis.*

The state with the lowest energy ($n = 1$) is called the ground state. It already follows from the uncertainty relations that the kinetic energy of the ground state must be different from zero [see problem

* The wave functions must then be normalized [see problem 2.2.1].

1.5.2]. Note that the number of nodes for the wave function inside the box equals the quantum number minus one.

We will now extend our considerations to a potential box in three dimensions. The potential energy is assumed to be zero in the region

$$-a/2<x<a/2, \ -b/2<y<b/2, \ -c/2<z<c/2,$$

and is infinitely high outside this region. The differential equation (2.1.17) inside the box will then be of the form

$$-\frac{\hbar^2}{2m}\left(\frac{\partial^2}{\partial x^2}+\frac{\partial^2}{\partial y^2}+\frac{\partial^2}{\partial z^2}\right)u\,(x, y, z) = Eu\,(x, y, z). \quad (2.2.11)$$

The variables can be separated by the trial function

$$u(x, y, z) = u_1(x)\, u_2(y)\, u_3(z), \quad (2.2.12)$$

which by insertion in (2.2.11) and division by $u_1\, u_2\, u_3$ gives

$$-\frac{\hbar^2}{2m}\left(\frac{u_1''}{u_1}+\frac{u_2''}{u_2}+\frac{u_3''}{u_3}\right) = E. \quad (2.2.13)$$

As the three ratios in the parentheses each depends only on one of the three variables, x, y and z, (2.2.13) can be replaced by the three differential equations

$$\begin{aligned}
-\frac{\hbar^2}{2m}u_1''(x) &= \mathscr{E}_1 u_1(x) \\
-\frac{\hbar^2}{2m}u_2''(y) &= \mathscr{E}_2 u_2(y) \\
-\frac{\hbar^2}{2m}u_3''(z) &= \mathscr{E}_3 u_3(z),
\end{aligned} \quad (2.2.14)$$

where the three separation constants must satisfy the equation

$$\mathscr{E}_1+\mathscr{E}_2+\mathscr{E}_3 = E. \quad (2.2.15)$$

As the wave function $u(x, y, z)$ must disappear on the boundary of the potential box, the three functions, u_1, u_2 and u_3, must disappear for $x = \pm a/2$, $y = \pm b/2$ and $z = \pm c/2$. The solution of each of the three differential equations (2.2.14) is obtained exactly as in the case of the one-dimensional potential box, and the separation constants act formally like energies. Thus the three-dimensional problem is reduced to the problem solved above. The differential equation (2.2.11) has acceptable solutions when the separation constants in (2.2.14) satisfy

$$\begin{aligned} \mathscr{E}_1 &= \frac{h^2}{8ma^2} n_1^2, \qquad n_1 = 1, 2, 3 \ldots \\ \mathscr{E}_2 &= \frac{h^2}{8mb^2} n_2^2, \qquad n_2 = 1, 2, 3 \ldots \\ \mathscr{E}_3 &= \frac{h^2}{8mc^2} n_3^2, \qquad n_3 = 1, 2, 3 \ldots \end{aligned} \tag{2.2.16}$$

Each of the possible stationary states is here characterized by three quantum numbers n_1, n_2 and n_3. The corresponding energy according to (2.2.15) and (2.2.16) is given by

$$E_{n_1 n_2 n_3} = \frac{h^2}{8m}\left(\frac{n_1^2}{a^2} + \frac{n_2^2}{b^2} + \frac{n_3^2}{c^2}\right), \tag{2.2.17}$$

while the wave function of the state is of the form

$$u(x, y, z) = A \begin{matrix} \sin \\ \cos \end{matrix} \left(\frac{n_1 \pi}{a} x\right) \begin{matrix} \sin \\ \cos \end{matrix} \left(\frac{n_2 \pi}{b} y\right) \begin{matrix} \sin \\ \cos \end{matrix} \left(\frac{n_3 \pi}{c} z\right), \tag{2.2.18}$$

where one must use sin or cos depending on whether n in the argument is even or odd.

The relative position of the energy levels (2.2.17) depends on the relative magnitude of a, b and c, and the energy spectrum can thus not be plotted in a universal diagram as in the case of the one-dimensional box. However, it is easily seen that the three-fold infinity of quantum numbers implies that the density of states (the number of levels in a given energy interval) increases with increasing energy [see problem 2.2.3]. Thus there is an obvious qualitative difference from the one-dimensional case, where the distance between the energy levels increases linearly with energy.

Moreover, it is important to note the difference between the number of states and the number of energy levels. Two states are different when they are described by linearly independent wave functions, which evidently is the case when the pertinent set of quantum number (n_1, n_2, n_3) are different. Usually one finds two different energy values from (2.2.17) for two different states. Only exceptionally —when the lengths a, b and c have special values—it happens that several different states have exactly the same energy. An energy level to which there belongs $m(>1)$ independent wave functions is said to be m times degenerate.

A strong degeneracy of the energy levels appears if the potential box is cubic, so that $a = b = c$. One then finds from (2.2.17) that

$$E_{n_1n_2n_3} = \frac{h^2}{8ma^2}({n_1}^2+{n_2}^2+{n_3}^2). \tag{2.2.19}$$

The ground state corresponding to $(n_1, n_2, n_3) = (1, 1, 1)$ is not degenerate, but already the next level is three times degenerate since the states with quantum numbers $(n_1, n_2, n_3) = (2, 1, 1)$, $(1, 2, 1)$ and $(1, 1, 2)$, all have the energy $6h^2/8ma^2$. Fig. 2.4 illustrates, on the

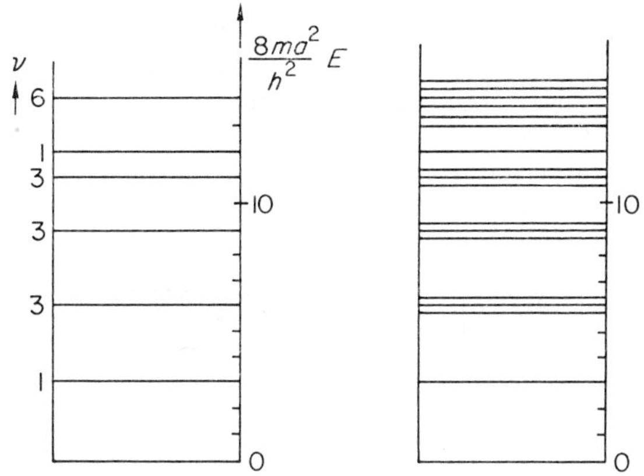

Fig. 2.4

left side, the lowest energy levels for a cubic potential box. The degree of degeneracy ν is indicated at each level. To the right is indicated how the energy spectrum may change when the box is deformed in such a way that the three lengths of the edges become slightly different. The former degenerate levels are split up, and the number of levels becomes the same as the number of states.

Problems

2.2.1 Show that the functions (2.2.9) become normalized when the constant A_n is chosen equal to $\sqrt{2/a}$ for all n.

2.2.2 Write down the quantum numbers for all the states which belong to the six lowest energy levels in a cubic potential box. Draw a level diagram for the analogous states in a potential box with the lengths a, b and c of the edges, where b is 10% smaller than a and c is 10% larger than a.

2.2.3 In the case of a potential box the distribution of states with large quantum numbers is easiest appreciated by imagining n_1, n_2 and n_3 to be the coordinates of points in an n_1, n_2, n_3 space. Show, that the points corresponding to states with energy lower than E fall inside an ellipsoidal surface, and thus find the approximate number, $N(E)$, of such states. Show next that the density of states at the energy E is

$$\rho_E = \frac{\mathrm{d}N}{\mathrm{d}E} = 4\pi \frac{m^{3/2}}{h^3} \Omega \sqrt{2E},$$

where Ω is the volume of the potential box.

2.2.4 The classical phase space volume is

$$\Gamma = \int \mathrm{d}p_x \mathrm{d}p_y \mathrm{d}p_z \mathrm{d}x \mathrm{d}y \mathrm{d}z.$$

From the result of problem 2.2.3, show that for particles in a potential box of volume Ω the number $N(E)$ of quantum states below the energy E is equal to the corresponding phase space volume divided by h^3 if $N(E) \gg 1$.

2.2.5 Show that for a one-dimensional potential box of width a, the momentum distribution in a stationary state with quantum number n is given by

$$P(p) = |f(p)|^2 = \frac{(a/\pi\hbar)(n\pi)^2}{[(pa/2\hbar)^2-(n\pi)^2]^2} \begin{cases} \sin^2 (pa/2\hbar), & \text{for } n \text{ even} \\ \cos^2 (pa/2\hbar), & \text{for } n \text{ odd.} \end{cases}$$

Examine the distribution for $n \to \infty$ and compare with the result of a classical description.

2.3 RECTANGULAR POTENTIAL WELL

We consider a particle in a potential which is similar to a potential box, with the exception that we now assume that the walls are not infinitely high. We confine the discussion to the one-dimensional case; a generalisation to three dimensions can be made in the same way as for the potential box.

The potential is assumed to be $\Phi(x) = 0$ for $-a/2 < x < a/2$, $\Phi(x) = \Phi_0 > 0$ for $x > a/2$ and $x < -a/2$. The differential equation (2.2.11) which determines the stationary states must be solved in each of the three regions which are shown in Fig. 2.5, and thereafter the solutions must be pieced together in such a way that u and u' become continuous at $x = \pm a/2$.

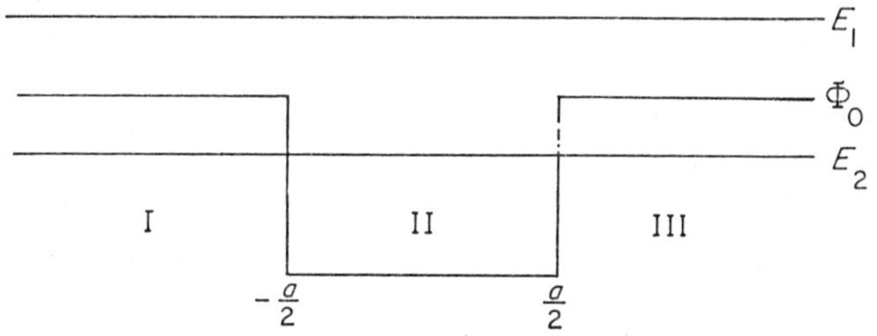

Fig. 2.5

The differential equations in the three regions are, in II,

$$u'' + \frac{2mE}{\hbar^2} u = 0; \tag{2.3.1}$$

and in I and III,

$$u'' + \frac{2m(E - \Phi_0)}{\hbar^2} = 0. \tag{2.3.2}$$

By the solution of (2.3.1) and (2.3.2) one must distinguish between the two cases $E > \Phi_0$ and $E < \Phi_0$, which are illustrated in the figure by the energies E_1, and E_2. According to classical mechanics the two cases correspond to situations where the particle is able to move everywhere on the x-axis or where it is bound to stay in the well.

For $E > \Phi_0$ the coefficients of u are positive in both (2.3.1) and (2.3.2), so that the solutions, $u(x)$, become oscillating in all three regions. By introducing the abbreviations

$$k = \left(\frac{2mE}{\hbar^2}\right)^{\frac{1}{2}}$$

and

$$k' = \left(\frac{2m(E - \Phi_0)}{\hbar^2}\right)^{\frac{1}{2}}, \tag{2.3.3}$$

one can write the general solution in three regions in the form

$$\begin{aligned} u_{\mathrm{I}}(x) &= A \sin k'x + B \cos k'x, \\ u_{\mathrm{II}}(x) &= C \sin kx + D \cos kx, \\ u_{\mathrm{III}}(x) &= E \sin k'x + F \cos k'x. \end{aligned} \tag{2.3.4}$$

These functions are finite for all x and can therefore be accepted as (improper) wave functions if the constants A, B, C, D, E and F can be adjusted in such a way that the functions satisfy the continuity conditions: $u_I(-a/2) = u_{II}(-a/2)$, $u_I'(-a/2) = u_{II}'(-a/2)$, $u_{II}(a/2) = u_{III}(a/2)$, and $u_{II}'(a/2) = u_{III}'(a/2)$. The continuity conditions give four restraints between the six constants, which means that one is always able to piece the solutions (2.3.4) together. Two of the constants can even be chosen arbitrarily, whereafter the rest can be determined from the continuity conditions. Such an explicit determination of the constants will not be carried through. We will only note that to each energy value $E > \Phi_0$ one can find stationary states for the particle, which can also be expressed by saying that the energy spectrum of the system is continuous for $E > \Phi_0$. We also notice that the twofold freedom in the choice of the constants in (2.3.4) means that each energy level in the continuous spectrum is doubly degenerate.

We will now consider the case $0 < E < \Phi_0$. The solutions in II become oscillating as before, while the solutions in the regions I and III become exponential functions of the type $\exp(\pm\beta x)$, where β is the real quantity

$$\beta = \left(\frac{2m(\Phi_0 - E)}{\hbar^2}\right)^{\frac{1}{2}} \tag{2.3.5}$$

As $u(x)$ must remain finite for $x \to \pm\infty$ we can only accept, in each of the regions I and III, one of the exponential functions, and the acceptable solutions in the three regions can be written in the form

$$\begin{aligned} u_I &= A \exp(\beta x), \\ u_{II} &= B \sin kx + C \cos kx, \\ u_{III} &= D \exp(-\beta x). \end{aligned} \tag{2.3.6}$$

Now we have to adjust only the four constants A, B, C and D according to the four continuity conditions, and the consequence is, as we shall see, that the solutions (2.3.6) can be pieced together only for special values of E.

From (2.3.6) one finds for the conditions at $x = -a/2$

$$u_I = u_{II} \to A \exp(-\beta a/2) = -B \sin(ka/2) + C \cos(ka/2) \tag{2.3.7}$$

$$u_I' = u_{II}' \to \beta A \exp(-\beta a/2) = k[B \cos(ka/2) + C \sin(ka/2)] \tag{2.3.8}$$

By division of (2.3.8) by (2.3.7) A is eliminated, and one gets

$$\beta = k\frac{B\cos(ka/2)+C\sin(ka/2)}{-B\sin(ka/2)+C\cos(ka/2)},$$

or

$$B[k\cos(ka/2)+\beta\sin(ka/2)]-C[\beta\cos(ka/2)-k\sin(ka/2)] = 0. \tag{2.3.9}$$

Writing the continuity conditions corresponding to (2.3.7) and (2.3.8) at $x = a/2$ one gets, after eliminating D,

$$B[k\cos(ka/2)+\beta\sin(ka/2)]+C[\beta\cos(ka/2)-k\sin(ka/2)] = 0, \tag{2.3.10}$$

which can also be deduced from (2.3.9) by performing the substitutions

$$\beta\to-\beta,\ a\to-a.$$

The equations (2.3.9), (2.3.10) have only non-vanishing solutions (B, C) when the coefficient of either B or C disappears, which means that k and β must satisfy one of the relations

$$\tan(ka/2) = \beta/k \tag{2.3.11}$$

$$-\cot(ka/2) = \beta/k. \tag{2.3.12}$$

When (2.3.11) or (2.3.12) are fulfilled, (2.3.9) and (2.3.10) are satisfied by $B = 0$, $C \neq 0$, and $C = 0$, $B \neq 0$, respectively. The wave function $u_{\text{II}}(x)$ in the well is therefore either proportional to $\cos kx$ [k satisfies (2.3.11)] or to $\sin kx$ [k satisfies (2.3.12)]. The ratios between the amplitude of $u_{\text{II}}(x)$ and the constants A and D can be determined, for example, from (2.3.7) and the corresponding continuity condition at $x = a/2$.

The possible energy values in the interval $0<E<\Phi_0$ are determined by (2.3.11) and (2.3.12) in connexion with (2.3.3) and (2.3.5) which express k and β in terms of E and the characteristic constants of the system, i.e. m, a and Φ_0. In order to get a survey of the solutions of the transcendental equations (2.3.11) and (2.3.12) we will proceed in the following manner. By means of (2.3.3) and (2.3.5), β can be eliminated and the right-hand sides of (2.3.11) and (2.3.12) can be written

$$\frac{\beta}{k} = \left(\frac{2m\Phi_0}{\hbar^2k^2}-1\right)^{\frac{1}{2}} = \left(\frac{\frac{1}{2}m\Phi_0a^2}{\hbar^2(ka/2)^2}-1\right)^{\frac{1}{2}} = \left(\frac{\kappa}{\xi^2}-1\right)^{\frac{1}{2}}. \tag{2.3.13}$$

In the last formulation β/k has been expressed in terms of two quantities

$$\kappa = \frac{m\Phi_0 a^2}{2\hbar^2}, \tag{2.3.14}$$

which contains the characteristic constants, and

$$\xi = \frac{ka}{2} = \left(\frac{mEa^2}{2\hbar^2}\right)^{\frac{1}{2}} = \left(\frac{\kappa E}{\Phi_0}\right)^{\frac{1}{2}}, \tag{2.3.15}$$

which for a given κ depends only upon the energy E measured in the unit Φ_0.

Using (2.3.13)–(2.3.15) one can then write (2.3.11) and (2.3.12) as

$$\left.\begin{matrix} \tan\xi \\ -\cot\xi \end{matrix}\right\} = \left(\frac{\kappa}{\xi^2} - 1\right)^{\frac{1}{2}}. \tag{2.3.16}$$

For a given κ, (2.3.16) is satisfied by a certain set of ξ–values, and the corresponding energy values can then be found from (2.3.15) in the form

$$E = \frac{1}{\kappa}\xi^2\Phi_0, \tag{2.3.17}$$

where ξ satisfies (2.3.16). One sees that for $E < \Phi_0$ the system has a discrete energy spectrum which depends on m, a and Φ_0 only through the dimensionless quantity κ.

Fig. 2.6 shows a graphic solution of (2.3.16) for $\kappa = 35$. In this case (2.3.16) is satisfied by four values which can be found approximately from the figure and afterwards be improved by substituting in (2.3.16). One finds

$$\xi_1 = 1{\cdot}34,\ \xi_2 = 2{\cdot}67,\ \xi_3 = 3{\cdot}97,\ \xi_4 = 5{\cdot}20,$$

which, according to (2.3.17), gives the four energy levels

$$E_1/\Phi_0 = 0{\cdot}051,\ E_2/\Phi_0 = 0{\cdot}204,\ E_3/\Phi_0 = 0{\cdot}45,\ E_4/\Phi_0 = 0{\cdot}77.$$

Fig. 2.7 shows the level diagram for $\kappa = 35$. The hatching for $E > \Phi_0$ indicates the continuous spectrum where the levels are overall dense. On the right are sketched the wave functions corresponding to the discrete energies and also their absolute squares.

One notices that the wave functions do not vanish exactly at

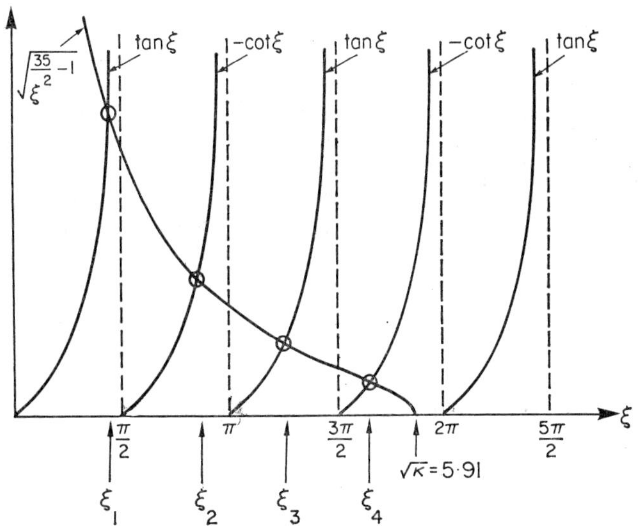

Fig. 2.6

$x = \pm a/2$, which means that there is a finite probability of finding the particle slightly outside the classically accessible region $-a/2 < x < a/2$. This probability apparently increases when the energy approaches Φ_0. When $E > \Phi_0$ the wave functions are improper, so that the particle—in accordance with the classical result—is not bound to the region of the well.

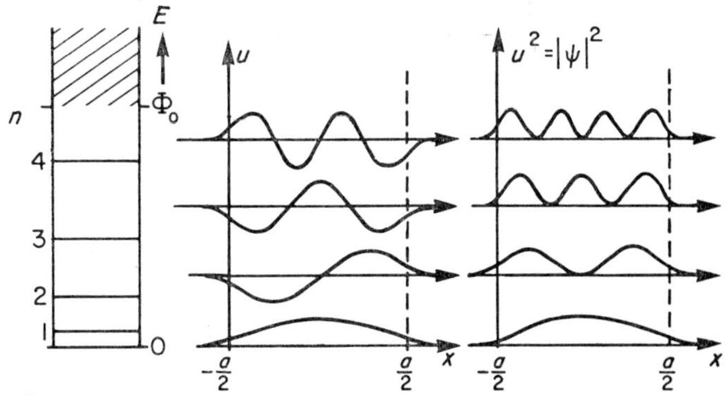

Fig. 2.7

Problems

2.3.1 Examine, for a one-dimensional potential well, those wave functions which belong to an energy in the continuous spectrum (consider, for example, even and odd functions of x). Show that the amplitude for $u(x)$ inside the well is smaller than or equal to the amplitude outside, and find under which circumstances they are equal.

2.3.2 Show, for a one-dimensional potential well, that the number, N, of discrete energy levels is finite. How does N depend on the quantity $\kappa = ma^2\Phi_0/(2\hbar^2)$? Find N for a well whose walls reach up to exactly the nth energy level in a box of the same width as the well.

2.3.3 Show that for the nth energy level in a one-dimensional well one has the approximate expressions

$$E_n \approx \frac{h^2}{8ma^2} n^2, \text{ when } E_n \ll \Phi_0,$$

and

$$E_n \approx \frac{h^2}{8ma^2}(n-1)^2, \text{ when } \Phi_0 - E_n \ll \Phi_0.$$

2.3.4 Find the energy spectrum for a particle in a one-dimensional potential well where one wall is infinitely high, as shown in Fig. 2.8.

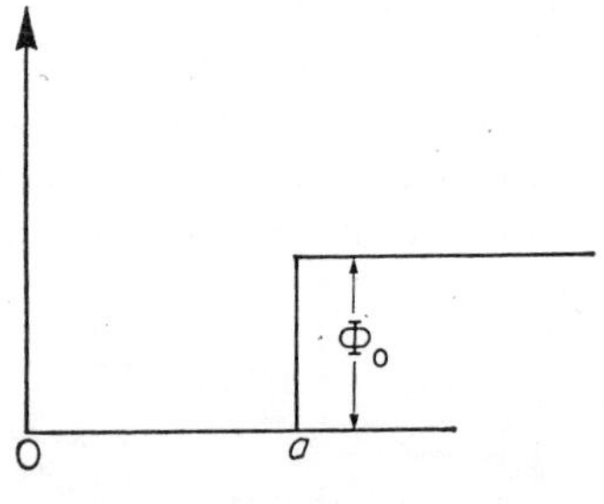

Fig. 2.8

2.4 THE HARMONIC OSCILLATOR

We consider a particle with mass m in a one-dimensional potential $\Phi(x) = kx^2/2$. According to classical mechanics a particle in this potential will perform harmonic oscillations with frequency

$$\omega = \sqrt{\frac{k}{m}}, \tag{2.4.1}$$

In order to examine the stationary states we must solve the differential equation (2.2.1), which in this case becomes

$$-\frac{\hbar^2}{2m}\frac{d^2u}{dx^2}+\tfrac{1}{2}kx^2u = Eu. \tag{2.4.2}$$

It is convenient to rewrite the differential equation in a dimensionless form. This can be obtained by introducing a new independent variable $\xi = \alpha x$, where

$$\alpha^4 = \frac{mk}{\hbar^2}. \tag{2.4.3}$$

The equation (2.4.2) may then be written

$$\frac{d^2u}{d\xi^2}+(\lambda-\xi^2)u = 0, \tag{2.4.4}$$

where λ is a dimensionless measure of the energy,

$$\lambda = \frac{2E}{\hbar\sqrt{k/m}} = \frac{2E}{\hbar\omega}. \tag{2.4.5}$$

In order to solve the differential equation (2.4.4) we will first try to determine the asymptotic behaviour of the solutions for $\xi\to\pm\infty$. For this purpose we write u as $u(\xi) = \exp[f(\xi)]$, i.e. $f(\xi) = \ln u(\xi)$, and seek the differential equation for $f(\xi)$. By substitution in (2.4.4) we get

$$f''+(f')^2+\lambda-\xi^2 = 0,$$

which for $|\xi|\to\infty$ has asymptotic solutions of the form $f(\xi)\sim\pm\xi^2/2$, corresponding to $u(\xi)\sim\exp(\xi^2/2)$ and $u(\xi)\sim\exp(-\xi^2/2)$. Only the latter of the two solutions corresponds to an acceptable wave function.

This investigation of the asymptotic behaviour leads us to seek solutions to (2.4.4) of the form

$$u(\xi) = \mathscr{H}(\xi)\exp(-\xi^2/2), \tag{2.4.6}$$

where the function $\mathscr{H}(\xi)$ for $\xi\to\pm\infty$ cannot tend towards infinity faster than $\exp(\xi^2/2)$.

Introducing (2.4.6) into (2.4.4) one finds that $\mathscr{H}(\xi)$ has to satisfy the differential equation

$$\mathscr{H}''-2\xi\mathscr{H}'+(\lambda-1)\mathscr{H} = 0. \tag{2.4.7}$$

We try to solve (2.4.7) by writing $\mathscr{H}(\xi)$ as a power series

$$\mathscr{H}(\xi) = \xi^s(a_0+a_1\xi+a_2\xi^2+\ldots) = \xi^s\sum_{m=0}^{\infty} a_m\xi^m \tag{2.4.8}$$

with

$$a_0\neq 0,\ s\geqslant 0.$$

The demand $a_0\neq 0$ means that s is the lowest power in the series, and cannot be negative since the wave function must be finite for $x = 0$.

We now introduce (2.4.8) into the left-hand side of (2.4.7) and collect terms with the same power of ξ. The condition for (2.4.8) to be a solution is then that the coefficient of every power of ξ must vanish. These conditions lead to the following set of equations:

$$\begin{aligned}
s(s-1)a_0 &= 0\\
(s+1)sa_1 &= 0\\
(s+2)(s+1)a_2-(2s+1-\lambda)a_0 &= 0\\
(s+3)(s+2)a_3-(2s+3-\lambda)a_1 &= 0\\
&\text{-----}\\
(s+m+2)(s+m+1)a_{m+2}-(2s+2m+1-\lambda)a_m &= 0\\
&\text{-----}
\end{aligned} \tag{2.4.9}$$

Since $a_0\neq 0$, it follows from the first equation that the lowest power in the series (2.4.8) must be $s = 0$ or $s = 1$. The second equation is automatically satisfied for $s = 0$. When $s = 1$ one must put $a_1 = 0$.

The coefficient a_0 may, for instance, be chosen to be 1. When s and a are chosen in accordance with the two first equations in (2.4.9) one is then able to calculate, by help of the recurrence relation, all the following coefficients a_m. For $s = 1$ we find $a_1 = 0$, which implies that all the coefficients a_m vanish for odd values of m. The series (2.4.8) then only contains odd powers of ξ and $\mathscr{H}(\xi)$ and becomes an odd function of ξ. For $s = 0$ one can obtain an even function of ξ by again choosing $a_1 = 0$. If a_1 is chosen to be non-zero the series will contain both odd and even powers.

We have thus seen that one is able to use the system of equations (2.4.9) to construct solutions to the differential equation (2.4.7). However, not all such solutions lead to acceptable wave functions. The wave function is only acceptable, as we shall see, when the series (2.4.8) terminates after a finite number of terms so that $\mathscr{H}(\xi)$ becomes a polynomial in ξ. If the series does not terminate we can determine the behaviour of $\mathscr{H}(\xi)$ for $\xi \to \pm\infty$ by examining the recurrence relation in (2.4.9) for $m \to \infty$. From (2.4.9) one finds that

$$\frac{a_{m+2}}{a_m} = \frac{2s+2m+1-\lambda}{(s+m+2)\,(s+m+1)} \to \frac{2}{m} \text{ for } m \to \infty. \qquad (2.4.10)$$

For the time being we will only consider the even terms in the series $\sum a_m \xi^m$. These must always be present, since $a_0 \neq 0$. Introducing $p = m/2$, the series of even terms can be written $\sum a_p' \xi^{2p}$ with a recurrence relation for a_p' which, according to (2.4.10) becomes $a_p'{}_{+1}/a_p' = 1/p$ for large p. If this recurrence relation were strictly valid for all p, the series of even terms would be

$$\sum_{p=0}^{\infty} \frac{1}{p!} (\xi^2)^p = \exp(\xi^2).$$

Now the simplified recursion formula is valid only for large p's, but as these terms dominate when $\xi \to \pm\infty$ we deduce that $\mathscr{H}(\xi)$ gets a contribution from terms with even m which behaves as $\exp(\xi^2)$. If, at the same time, an infinite series of terms with odd m occurs in (2.4.8), one can show in the same manner as above that their contribution to $\mathscr{H}(\xi)$ also tends towards infinity like $\exp(\xi^2)$. Comparing with (2.4.6) one sees that such functions do not lead to acceptable wave functions.

Consequently the series (2.4.8) must terminate, and according to the recurrence relation (2.4.10) this can only happen if λ is an odd integer. We put $\lambda = 2n+1$ (where n is an integer) and we see from (2.4.10) that $2s+2m+1-\lambda$ becomes zero for $m = n-s$. Thus all coefficients $a_{n-s+2}, a_{n-s+4}, \ldots$ vanish. As we have seen, terms with even m always occur in the series, and we demand that m be even. The series of terms with odd m in this case does not terminate. It thus follows that we must always choose $a_1 = 0$, so that all the odd coefficients automatically vanish.

The values of λ for which (2.4.4) has acceptable solutions are, according to the above,

$$\lambda_n = 2n+1, \qquad n = 0, 1, 2, \ldots \; . \tag{2.4.11}$$

The energy levels of the oscillator are determined by (2.4.11) and (2.4.5), leading to

$$E_n = (n+\tfrac{1}{2})\hbar\omega, \qquad n = 0, 1, 2, \ldots, \tag{2.4.12}$$

where ω is the classical frequency (2.4.1) of the oscillator.

The wave functions for the stationary states with energies (2.4.12) are of the form (2.4.6), where, according to the above, $\mathscr{H}(\xi)$ is a polynomial of degree n. These polynomials can, except for a constant factor, be determined unambiguously by means of the recurrence relation in (2.4.9). We must here put $a_1 = 0$, $\lambda = 2n+1$, and choose $s = 0$ or $s = 1$ for even or odd n respectively (since $n-s$ has to be even).

It may be shown that the polynomial which enters the wave function for the nth state can be written explicitly as

$$H_n(\xi) = (-1)^n \exp(\xi^2) \frac{\mathrm{d}^n}{\mathrm{d}\xi^n} \exp(-\xi^2), \tag{2.4.13}$$

where we have made a choice for the undetermined coefficient a_0'. The function (2.4.13) is called a Hermite polynomial of degree n. With the definition (2.4.13) the wave function corresponding to the energy E_n can be written as

$$u(\xi) = A_n H_n(\xi) \exp(-\xi^2/2), \tag{2.4.14}$$

with $\xi = \alpha x$, where α given by (2.4.3). The constant A_n can be adjusted to make u normalized. The normalization factor can be proved to be

$$A_n = \left(\frac{\alpha}{\pi^{\frac{1}{2}} 2^n n!}\right)^{\frac{1}{2}}.$$

The lowest Hermite polynomials are found to be

$$H_0(\xi) = 1, \quad H_1(\xi) = 2\xi, \quad H_2(\xi) = 4\xi^2 - 2,$$

$$H_3(\xi) = 8\xi^3 - 12\xi, \quad H_4(\xi) = 16\xi^4 - 48\xi^2 + 12.$$

The upper part of Fig. 2.9 shows the level diagram for the oscillator and the normalized wave functions for the four lowest states. The

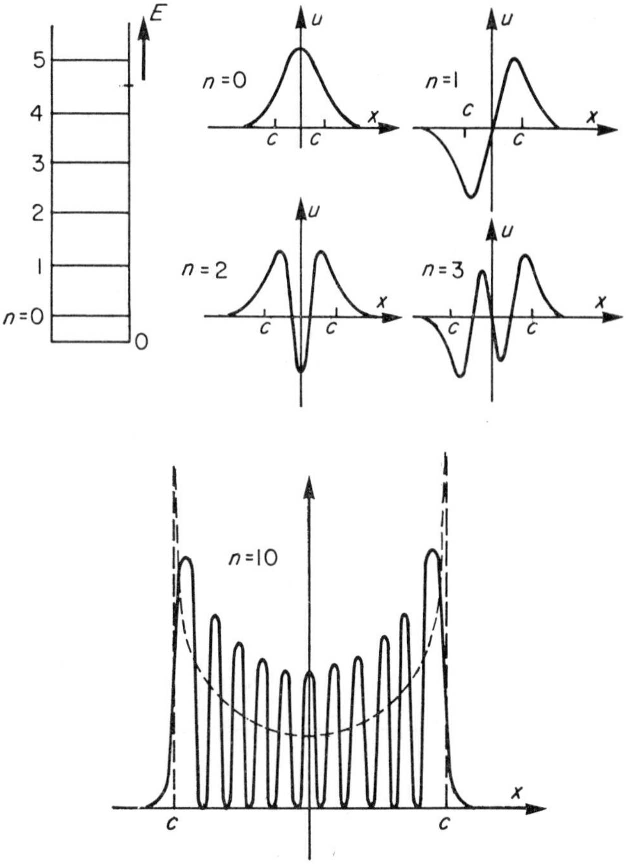

Fig. 2.9

intervals c–c indicate the regions which are accessible for the particle in classical mechanics. Below, one sees the probability distribution on the x-axis for a state with $n = 10$. For comparison the classical probability of finding the particle in an interval dx at x is also plotted (broken line). This probability is proportional to the time spent in the interval by the particle in the classical description, so that the classical probability has a minimum at the centre, while it tends towards infinity at the turning points. For states with large quantum numbers the quantum mechanical description yields a probability distribution which in the mean coincides with the classical distribution.

Problems

2.4.1 A particle of mass 1 g performs harmonic oscillations with an amplitude of 1 cm and a period of 1 sec. Find the magnitude of the quantum number which enters in a quantum mechanical description.

Find the distance between the energy levels for a harmonic oscillator with a period of 10^{-15} sec.

2.4.2 Discuss the energy spectrum for a three-dimensional harmonic oscillator with potential energy $\Phi(r) = (1/2)\,kr^2$. Determine the degree of degeneracy for the four lowest energy levels.

2.4.3 Show from (2.4.13) that the Hermite polynomials satisfy the recurrence relations

$$H_n' = 2nH_{n-1} \text{ and } H_{n+1} = 2\xi H_n - 2nH_{n-1};$$

show then that H_n satisfies (2.4.7) with $\lambda = 2n+1$.

2.4.4 A particle with mass m in a potential $\frac{1}{2}kx^2$ is not necessarily in a stationary state. Express the mean value of the energy, E, in terms of the four quantities

$$\langle x\rangle,\ \langle p\rangle,\ \sigma_x \text{ and } \sigma_p.$$

Thus determine the lowest value of E which is consistent with the uncertainty relations, and show that this minimum value can actually be realized. Interpret the result by comparing with problem 1.5.2.

2.5 SUPERPOSITION OF STATIONARY STATES

In the last sections we found in some specific cases the wave functions for stationary states. We will now discuss some general properties of wave functions for stationary states and next we will investigate superpositions of such stationary wave functions. These superpositions describe states which change with time and which cannot be ascribed a definite energy. We thus consider the solutions $u_E(\mathbf{r})$ of the eigenvalue equation

$$\hat{H}u_E(\mathbf{r}) = Eu_E(\mathbf{r}), \tag{2.5.1}$$

corresponding to the energy E, and note first of all that the eigenvalues E always are real, in accordance with their physical significance. In order to show this, write the energy as an expectation value

$$E = \int u_E{}^*\hat{H}u_E\mathrm{d}\tau. \tag{2.5.2}$$

Utilizing the fact that $\hat{H}$ is a Hermitian operator [see (1.6.20)] one obtains

$$\int u_E{}^*\hat{H}u_E\mathrm{d}\tau = \int u_E(\hat{H}u_E)^*\mathrm{d}\tau = E^*,$$

which together with (2.5.2) at once yields

$$E = E^*. \tag{2.5.3}$$

One can, of course, generalize this argument to other operators, and it thus follows that all Hermitian operators have real expectation values and eigenvalues. This is the reason that one must generally expect that operators which, like $\hat{\mathbf{p}}$ and $\hat{H}$, have a physical significance, must be Hermitian.

We will now consider two solutions u_{E_1} and u_{E_2} of (2.5.1), which belong to the eigenvalues E_1 and E_2 respectively, i.e.:

$$\hat{H}u_{E_1} = E_1 u_{E_1}, \tag{2.5.4}$$

and

$$\hat{H}u_{E_2} = E_2 u_{E_2}. \tag{2.5.5}$$

As above [(2.5.2) and (2.5.3)] we will calculate the 'matrix element' $\int u_{E_1}\hat{H}u_{E_2}\mathrm{d}\tau$ in two ways. From (2.5.5) we find directly

$$\int u_{E_1}^*\hat{H}u_{E_2}\mathrm{d}\tau = E_2\int u_{E_1}^* u_{E_2}\mathrm{d}\tau. \tag{2.5.6}$$

On the other hand, since $\hat{H}$ is Hermitian, we find from (2.5.4) that

$$\int u_{E_1}^*\hat{H}u_{E_2}\mathrm{d}\tau = \int u_{E_2}(\hat{H}u_{E_1})^*\mathrm{d}\tau = E_1{}^*\int u_{E_1}^* u_{E_2}\mathrm{d}\tau. \tag{2.5.7}$$

Finally, subtracting (2.5.7) from (2.5.6) and using (2.5.3) we get

$$(E_2 - E_1)\int u_{E_1}^* u_{E_2}\mathrm{d}\tau = 0. \tag{2.5.8}$$

From this equation we may now conclude that the integral on the right-hand side of (2.5.6) vanishes when E_1 and E_2 are different, i.e.

$$\int u_{E_1}^* u_{E_2}\mathrm{d}\tau = 0, \quad \text{for } E_1 \neq E_2. \tag{2.5.9}$$

Two wave functions which satisfy (2.5.9) are said to be orthogonal, and we have thus shown that two proper eigenfunctions belonging to stationary states with different energy are mutually orthogonal.

One cannot use the same argument for improper wave functions, which belong to the levels in a continuous spectrum. For such functions the integral (2.5.9) does not converge in the usual sense. However, this is not due to any 'profound' difference between states described by proper and improper wave functions. It is possible to adapt the mathematics in such a way that (2.5.9) holds also for improper wave functions. We will here, however, use another procedure in which we make an imperceptible change in the physical system in order that all wave functions may be treated in the same fashion. This can be achieved by imagining the whole system confined in a large potential box, on the boundary of which the wave functions must vanish. For the system in the box all wave functions become proper, and the continuous spectrum is substituted by discrete energy levels which, by choosing the box sufficiently large, can be brought to lie arbitrarily close. It is obvious that one can always choose the box so large that it does not have any physical influence on a finite system*. One is thus able freely to introduce or leave out the box according to whether it is convenient or not for the mathematical treatment of the problem.

The orthogonality of the stationary wave functions allows a simple determination of the energy distribution in a state whose wave function is a superposition of stationary wave functions. Let $u_{E_1}, u_{E_2}, \ldots u_{E_n}$ be normalized solutions to the stationary Schrödinger equation, corresponding to energies which are all different. According to (2.5.9) and the normalization of the wave functions we then have

$$\int u_{E_m}^* u_{E_l} \mathrm{d}\tau = \delta_{ml} = \begin{cases} 1 \text{ for } m = l \\ 0 \text{ for } m \neq l. \end{cases} \tag{2.5.10}$$

We now consider a wave function which at the time $t = 0$ is a superposition of the u's:

$$\psi(\mathbf{r}, 0) = C_1 u_{E_1} + C_2 u_{E_2} + C_3 u_{E_3} + \ldots = \sum_{m=1}^{n} C_m u_{E_m}(\mathbf{r}). \tag{2.5.11}$$

In order to have ψ normalized the constants C_m must satisfy

$$\int |\psi|^2 \mathrm{d}\tau = \sum_{lm} C_l^* C_m \int u_{E_l}^* u_{E_m} \mathrm{d}\tau = \sum_{lm} C_l^* C_m \delta_{lm} = \sum_m |C_m|^2 = 1, \tag{2.5.12}$$

where we have made use of (2.5.10).

Since all the u's are eigenfunctions for the Hamiltonian, one finds from (2.5.11) by letting $\hat{H}$ act s times on ψ

$$\hat{H}^s \psi = C_1 E_1{}^s u_{E_1} + C_2 E_2{}^s u_{E_2} + \cdots \quad . \tag{2.5.13}$$

* Note that it is only under this assumption that it is at all possible to treat, for instance, an atom as an isolated system.

Multiplication from the left by ψ^* and integration over the entire space yields as the mean value of the s'th power of the energy

$$\langle E^s \rangle = \int \psi^* \hat{H}^s \psi \mathrm{d}\tau = \sum_{lm} C_l^* C_m E_m{}^s \int u_{E_l}^* u_{E_m} \mathrm{d}\tau, \quad (2.5.14)$$

or—using (2.5.10)—

$$\langle E^s \rangle = |C_1|^2 E_1{}^s + |C_2|^2 E_2{}^s + \ldots = \sum_{m=1}^{n} |C_m|^2 E_m{}^s. \quad (2.5.15)$$

From (2.5.15) and (2.5.12) we see that the quantities $|C_1|^2$, $|C_2|^2, \ldots$ $|C_n|^2$ must be identified with the probabilities $P_1, P_2, \ldots, P_n$ of finding the particle with the well-defined energies $E_1, E_2 \ldots, E_n$ in the state (2.5.11). This means that in the state (2.5.11) an accurate measurement of the energy of the particle will lead to one of the values $E_1, E_2, \ldots E_n$. It cannot be predicted with certainty which one of these will be found but one is able to state the probability P_l of finding the lth energy value.

In order to calculate P_l it is not necessary to write down the wave function explicitly as a superposition of stationary wave functions. By multiplying (2.5.11) by u_l^* and integrating over space, one obtains by using (2.5.10)

$$C_l = \int u_{E_l}^*(\mathbf{r}) \psi(\mathbf{r},0) \mathrm{d}\tau, \quad (2.5.16)$$

the absolute square of which yields the probability P_l.

Above, we have only considered a special superpostion of stationary wave functions beloning to different energies where the wave functions were automatically orthogonal to each other. If an energy level is degenerate so that several linearly independent wave functions belong to the same energy value, one can easily verify that it is possible to form mutually orthogonal linear combinations of these functions. One is therefore able to associate all the stationary states (possibly after enclosing the system in a potential box) with wave functions which form an infinite set of orthogonal functions. One assumes that this set of functions is complete in the sense that any function which can be accepted as a wave function can be expanded using the solutions to the stationary Schrödinger equation.

From this it follows that it should be possible to write any wave function as a (possibly infinite) series of the form (2.5.11). If the system has a continuous spectrum and one does not wish to introduce a potential box, one must write the contribution to (2.5.11) from the continuous spectrum as an integral and the expansion of the wave function on the eigenfunctions u_E will then have the form

$$\psi(\mathbf{r}, 0) = \sum_{\text{discrete}} C_m u_{E_m}(\mathbf{r}) + \int_{\text{continuous}} C(E) u_E(\mathbf{r}) \mathrm{d}E. \tag{2.5.17}$$

The advantage of separating the wave function (at time $t = 0$) into stationary wave functions is that one can then easily find the development in time of the state. If each of the functions $u_E(\mathbf{r})$ in (2.5.17) is multiplied by the time factor given by (2.1.10), then the product, $\psi_E(\mathbf{r}, t) = u_E(\mathbf{r}) \exp(-\mathrm{i}Et/\hbar)$ will be a solution of the time-dependent Schrödinger equation (2.1.1), and since this equation is linear in ψ any linear combination of solutions will again be a solution.

Consequently the wave function

$$\psi(\mathbf{r}, t) = \sum_m C_m u_{E_m}(\mathbf{r}) \exp\left(-\frac{i}{\hbar} Et\right) + \int C(E) u_E(\mathbf{r}) \exp\left(-\frac{i}{\hbar} Et\right) \mathrm{d}E \tag{2.5.18}$$

satisfies the Schrödinger equation and represents the development of the state (2.5.17) in time.

The calculation of how a given state will change in time can thus be reduced to the determination of the energy spectrum and the eigenfunctions $u_E(\mathbf{r})$ for the potential in question. In the following chapter we will make extensive use of this method when examining the motion of particles in various potentials in terms of the solutions of the stationary Schrödinger equation. As a conclusion to the present chapter we shall merely illustrate the above discussion by considering some simple examples.

Let $\psi(\mathbf{r}, t)$ be a superposition of only two stationary wave functions, so that

$$\psi(\mathbf{r}, t) = C_1 u_{E_1} \exp\left(-\frac{i}{\hbar} E_1 t\right) + C_2 u_{E_2} \exp\left(-\frac{i}{\hbar} E_2 t\right). \tag{2.5.19}$$

If we disregard the trivial possibility $E_1 = E_2$, the energy in the state (2.5.19) is undetermined, and $|C_1|^2$ and $|C_2|^2$ indicate the probabilities that the energy of the particle is E_1 or E_2, respectively. The probability density in space is given by

$$|\psi|^2 = |C_1|^2 |u_{E_1}(\mathbf{r})|^2 + |C_2|^2 |u_{E_2}(\mathbf{r})|^2 + 2\mathcal{R}e\left[C_1 C_2{}^* u_{E_1}(\mathbf{r})\, u_{E_2}^*(\mathbf{r}) \exp\left(\frac{i}{\hbar}(E_2 - E_1)t\right)\right], \tag{2.5.20}$$

which is seen to be periodic in time with the period $T = 2\pi\hbar/(E_2 - E_1)$. By superposition of more than two stationary states the probability density will contain a sum of terms which usually have incommensurable periods and $|\psi|^2$ is then no longer a periodic function of time. The classical motion of a particle bound in a one-dimensional potential is, however, periodic and the probability density in a quantum mechanical description of such a classical motion must therefore, at least to a high accuracy, be periodic in time. How such an approximate periodicity can result is shown for a potential box in problem 2.5.4, where one makes a superposition of close-lying states with large quantum numbers n. The probability density for such a state can be regarded as periodic during a limited time and the period for $n \to \infty$ will be equal to the period in the classical motion of the particle.

As we shall see, the harmonic oscillator has the special feature that the probability density in an arbitrary state is periodic in time, and the period is the same as the classical period $T_{cl} = 2\pi/\omega$, where $\omega = \sqrt{k/m}$. Thus an arbitrary wave function for the oscillator can be written as

$$\psi(x, t) = \sum_n C_n u_{E_n}(x) \exp\left(-\frac{i}{\hbar} E_n t\right), \tag{2.5.21}$$

where the functions u_{E_n} are given by (2.4.14) and the corresponding energies are given by

$$E_n = (n+\tfrac{1}{2})\hbar\omega. \tag{2.5.22}$$

The phase in the time factor of the nth term is thus $E_n t/\hbar = (n+\frac{1}{2})\omega t$, and all the terms will therefore change their sign when the time increases by $T_{cl} = 2\pi/\omega$. Hence, $\psi(x, t) = -\,\psi(x, t+T_{cl})$, and the probability density is therefore periodic in time with the period T_{cl}. If one uses furthermore the parity of the stationary wave functions then it is possible also to draw conclusions about the change of the state during half a period. When the time increases by $T_{cl}/2$, the phases in all the time factors increase by $n\pi+\pi/2$ and the terms are thus multiplied by $-i$ or $+i$ according to whether n is even or odd. But, since we found in 2.4 that the functions $u_n(x)$ in (2.5.21) have even parity for even n and odd parity for odd n, we conclude that $\psi(x, t) = -\,i\psi(-x, t+T_{cl})$, which means that after a period $T_{cl}/2$ the probability density is reflected in $x = 0$. Fig. 2.10 indicates how the probability density for an arbitrary state of a harmonic oscillator moves back and forth with the classical period. In order to follow the motion in more detail one must, for a special choice of coefficients in (2.5.19), calculate the probability density explicitly.

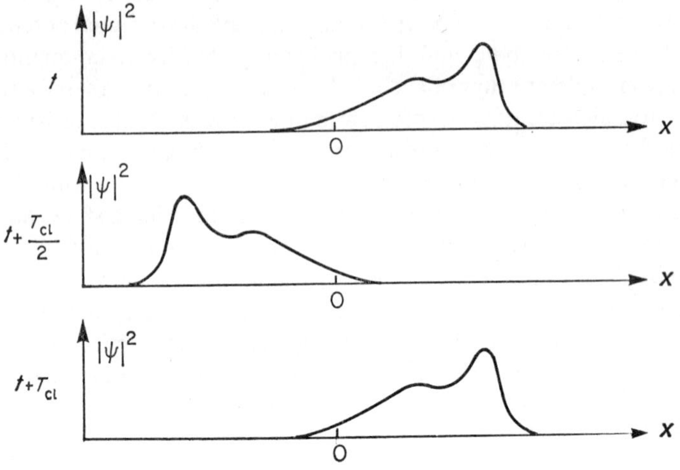

Fig. 2.10

The properties of the stationary states, which we have discussed in this section, are solely founded on the fact that the wave functions $u_{E_n}(\mathbf{r})$ are eigenfunctions of a Hermitian operator. We can without any further comments conclude that the eigenfunctions of operators, corresponding to other physical quantities than the energy, also show similar properties. As an example we may consider the one-dimensional case of eigenfunctions to the momentum operator $\hat{p}_x$:

$$\hat{p}_x u_p(x) = p u_p(x). \tag{2.5.23}$$

The assumption that the eigenfunctions $u_p(x)$ form a complete orthogonal system is in this case known to be correct from the integral theorem of Fourier, since the solution to (2.5.23) is

$$u_p(x) = \alpha \exp\left(\frac{i}{\hbar} px\right), \tag{2.5.24}$$

where α is an arbitrary constant. When, in 1.4, we expanded an arbitrary wave packet on the functions (2.5.24), this was an expansion which corresponded closely to the expansion in the last term of

(2.5.17)–(2.5.18). Since the wave functions (2.5.23) describe states with well-defined momenta [see (2.1.16)], the interpretation of the coefficient of expansion $f(\mathbf{p})$ in (1.4.9) is completely analogous to the interpretation of the function $C(E)$ in (2.5.16).

Problems

2.5.1 Show that any one-dimensional wave function which vanishes for $x = \pm a/2$ can be written in the form

$$\sum_n C_n u_{E_n}(x),$$

where the u's are the stationary wave functions (2.2.9) for the potential box.

2.5.2 Show, by the integral theorem of Fourier, that any one-dimensional proper wave function can be represented by

$$\int_0^\infty C(E) u_E(x) \mathrm{d}E,$$

where $u_E(x)$ is a solution, corresponding to the energy E, of the stationary Schrödinger equation for a free particle on the x-axis.

2.5.3 Consider, in a one-dimensional potential box, a superposition of stationary wave functions corresponding to quantum numbers between $n-p$ and $n+p$. Show that the quantity $|E_n - \langle E \rangle|/E_n$ and the relative fluctuation in energy, σ_E/E_n, are both smaller than $4p/n$. This indicates that for $n \to \infty$ and p fixed one can consider the energy of the state as sharply and adequately represented by the value E_n.

Show that the probability density on the x-axis depends on the time through a sum of terms which vary harmonically and whose phases for $p \ll n$ can be written approximately as: t times an integer times a constant, common to all terms. In this approximation the probability density varies periodically in time. Find the period and show that it is identical to the classical one for a particle with energy E_n. Estimate the time, measured in periods, during which the periodicity will be maintained.

2.5.4 Let $u_n(x)$ and $u_m(x)$ denote the stationary wave functions for a one-dimensional harmonic oscillator where n and m are the quantum numbers. Use the result of problem 2.4.3 and show that the integral $\int u_n^* x u_m \mathrm{d}x$ is zero, unless $n = m \pm 1$.

Then show that for an arbitrary state of a harmonic oscillator the mean value $\langle x \rangle$ varies harmonically in time with the cyclic frequency ω.

2.6 APPROXIMATION METHODS

In the preceding sections we have found the stationary states for a particle in some special cases. However, only in a few potentials $\Phi(\mathbf{r})$ is it thus possible to solve explicitly the eigenvalue problem (2.1.11), and hence it is very important for practical purposes to introduce systematic methods by which the problem may be solved by successive approximations. The present section deals with two such methods, i.e. the Ritz variation method and the perturbation calculation.

The former method is particularly suited for the determination of the ground state of a system. It is based on the simple observation that the expression (2.5.15) for the mean value of the energy is minimal if the system is in its ground state. Thus if we write down $\langle E\rangle$ for an arbitrary state described by the normalizable wave function $u(\mathbf{r})$, we find, according to (2.5.15),

$$\langle E\rangle = \frac{\int u^*(\mathbf{r})\hat{H}u(\mathbf{r})\mathrm{d}\tau}{\int |u(\mathbf{r})|^2\mathrm{d}\tau}$$

$$= |C_1|^2E_1+|C_2|^2E_2+\ldots\ldots \tag{2.6.1}$$

where C_n is the expansion coefficient for the state u on the stationary states u_{E_n}. Since

$$\sum_n |C_n|^2 = 1$$

we see directly that $\langle E\rangle$ assumes its lowest value, which is the ground state energy E_1, only if $|C_1| = 1$, i.e. if $u(\mathbf{r}) = u_{E_n}(\mathbf{r})$.

This result may also be derived from (2.6.1) by calculating the change in $\langle E\rangle$ for small changes in u. We introduce a wave function $\tilde{u}(\mathbf{r})$ which everywhere deviates only slightly from $u(\mathbf{r})$, and calculate the difference between $\langle E\rangle$ for the state $\tilde{u}(\mathbf{r})$ and for the state $u(\mathbf{r})$ to first order in the difference $\delta u(\mathbf{r}) = \tilde{u}(\mathbf{r})-u(\mathbf{r})$. Since we have seen in 2.1 that the wave functions for stationary states may be chosen to be real functions, we assume that both u and $\tilde{u}$ are real and find

$$\delta\langle E\rangle = \langle E\rangle_{\tilde{u}}-\langle E\rangle_u$$

$$= \frac{\int(\delta u\hat{H}u+u\hat{H}\delta u)\mathrm{d}\tau}{\int u^2\mathrm{d}\tau} - \frac{2\int u\hat{H}u\mathrm{d}\tau\int(u\delta u)\mathrm{d}\tau}{(\int u^2\mathrm{d}\tau)^2}$$

$$= 2\frac{\int\delta u(\hat{H}u-\langle E\rangle_u u)\mathrm{d}\tau}{\int u^2\mathrm{d}\tau} \tag{2.6.2}$$

where we have used the Hermiticity of $\hat{H}$ (1.6.20).

If u represents the wave function belonging to a minimum for $\langle E\rangle$, one must find

$$\delta\langle E\rangle = 0, \tag{2.6.3}$$

for all small variations δu, and from this we conclude that the integrand in (2.6.2) is zero, i.e.

$$\hat{H}u = \langle E\rangle u. \tag{2.6.4}$$

Thus the wave function u is an eigenfunction of $\hat{H}$ with the eigenvalue $\langle E\rangle$. According to the previous remarks, there exists then for $u = u_{E_1}$ an absolute minimum of $\langle E\rangle$ for variations of u away from u_{E_1}.

The above variation principle may be used directly in an approximate determination of the ground state of a system. In this case one tries to choose a wave function, $u[\mathbf{r},\alpha,\beta,\ldots]$ which depends on a number of parameters $\alpha, \beta, \ldots$. Among the class of functions of $\mathbf{r}$ formed by varying $\alpha, \beta, \ldots$ one may determine the one which best simulates the true eigenfunction of the ground state by determining the minimum value of

$$f(\alpha,\beta,\ldots) = \langle E\rangle = \frac{\int u(\mathbf{r},\alpha,\beta,\ldots)\hat{H}u(\mathbf{r},\alpha,\beta,\ldots)\mathrm{d}\tau}{\int u(\mathbf{r},\alpha,\beta,\ldots)^2\mathrm{d}\tau}. \tag{2.6.5}$$

A variation of the parameters leads to

$$\delta\langle E\rangle = \frac{\partial f}{\partial\alpha}\,\delta\alpha + \frac{\partial f}{\partial\beta}\,\delta\beta + \ldots. \tag{2.6.6}$$

The minimum in $\langle E\rangle$ is thus a solution of the set of equations

$$\frac{\partial f}{\partial\alpha} = \frac{\partial f}{\partial\beta} = \ldots = 0. \tag{2.6.7}$$

These equations usually have several solutions and we must choose the one that leads to the lowest value of $f = \langle E\rangle$.

The minimum value of $\langle E\rangle$ is an approximation to the energy, E_1, in the ground state. The accuracy of this approximation depends on the initial choice of ansatz functions u. However, even with a mediocre estimate, the deviation of $\langle E\rangle$ from E_1 is still of 2nd order in the deviation of u from u_1, since one is close to a minimum in $\langle E\rangle$.

As an example of the use of the Ritz variation method we indicate an approximation to the ground state of the hydrogen atom. In this case the Hamiltonian operator is

$$\hat{H} = -\frac{\hbar^2}{2m}\Delta - \frac{e^2}{r}, \tag{2.6.8}$$

r being the distance of the electron from the proton. We know that in the ground state the electron is close to the nucleus; let us therefore assume the wave function to be of the type

$$u = \exp(-r^2/\alpha^2), \tag{2.6.9}$$

α being a parameter indicating the 'radius' of the atom. We work out the normalizing integral to be

$$\int |u^2| \mathrm{d}\tau = \left(\frac{\pi}{2}\right)^{3/2} \alpha^3. \tag{2.6.10}$$

Since

$$\Delta u = \exp(-r^2/\alpha^2)\, \frac{4r^2 - 6\alpha^2}{\alpha^4}, \tag{2.6.11}$$

we find by simple calculation that

$$\langle E \rangle = 4\left(\frac{8}{\pi}\right)^{\frac{1}{2}} \left(\frac{3}{8}\frac{\hbar^2}{2m\alpha^2}\left(\frac{\pi}{2}\right)^{\frac{1}{2}} - \frac{e^2}{4}\frac{1}{\alpha}\right). \tag{2.6.12}$$

The best value of the parameter α is determined by the equation

$$\frac{\partial \langle E \rangle}{\partial \alpha} = 4\left(\frac{8}{\pi}\right)^{\frac{1}{2}} \left(-\frac{3}{4}\frac{\hbar^2}{2m\alpha^3}\left(\frac{\pi}{2}\right)^{\frac{1}{2}} + \frac{e^2}{4\alpha^2}\right) = 0,$$

i.e.

$$\alpha = \frac{3}{2}\left(\frac{\pi}{2}\right)^{\frac{1}{2}} a_0 \tag{2.6.13}$$

where $a_0 = \hbar^2/me^2$ is the so-called Bohr radius. The corresponding energy is

$$\langle E \rangle_{\min} = -\frac{4}{3}\frac{me^4}{\pi\hbar^2} = -11{\cdot}6 \text{ eV}. \tag{2.6.14}$$

We note that the wave function (2.6.9) gives only a poor approximation to the wave function $u = (\pi a_0{}^3)^{-\frac{1}{2}} \exp(-r/a_0)$ [see (4.3.16)] for the ground state of the hydrogen atom. Still, the energy value is rather close to the correct value of $-13{\cdot}6$ eV.

When choosing the ansatz function in the variation method it is a great help if one knows the eigenfunction of a system, the Hamiltonian of which is similar to the one in question. Under such circumstances one may, however, introduce a more systematic approximation procedure. Suppose that the Hamilton operator is of the form

$$\hat{H} = \hat{H}_0 + \hat{\Omega}, \tag{2.6.15}$$

where the solutions to the eigenvalue problem for $\hat{H}_0$ is known, i.e. we know all the eigen functions $v_n(\mathbf{r})$ and the corresponding energies $E_n^{(0)}$ satisfying the equation

$$\hat{H}_0 v_n(\mathbf{r}) = E_n^{(0)} v_n(\mathbf{r}). \tag{2.6.16}$$

It should be noted that if the spectrum is degenerate, different values of n can correspond to the same energy eigenvalue. We seek solutions to the eigenvalue problem

$$(\hat{H}_0 + \hat{\Omega}) u(\mathbf{r}) = E u(\mathbf{r}), \tag{2.6.17}$$

where we assume the contribution $\hat{\Omega}$, the so-called perturbation energy, to be small. Under these circumstances it is convenient to write the unknown eigenfunction $u(\mathbf{r})$ as a series expansion in the complete orthogonal system $v_m(\mathbf{r})$:

$$u(\mathbf{r}) = \sum_m a_m v_m(\mathbf{r}). \tag{2.6.18}$$

We introduce this in (2.6.17), multiply the resulting equation by $v_n^*(\mathbf{r})$, and integrate throughout space. This leads to the following system of equations for the determination of a_n:

$$(E - E_n^{(0)}) a_n = \sum_m \Omega_{nm} a_m, \tag{2.6.19}$$

where

$$\Omega_{nm} = \int v_m^*(\mathbf{r}) \hat{\Omega} v_n(\mathbf{r}) d\tau \tag{2.6.20}$$

is called the matrix element of $\hat{\Omega}$ between the states m and n.

We now try to expand the solution of (2.6.19) in power of the small quantities $\hat{\Omega}$. To achieve this, we write $\hat{\Omega}$ in the form

$$\hat{\Omega} = \lambda \hat{\omega}, \tag{2.6.21}$$

where λ is a small, real number. We can then expand E in powers of λ:

$$E_l = E_l^{(0)} + \lambda E_l^{(1)} + \lambda^2 E_l^{(2)} + \dots \, . \tag{2.6.22}$$

For $\lambda = 0$, E must be equal to one of the eigenvalues $E_l^{(0)}$, and we have therefore given E the suffix l. Similarly we may expand a_n in powers of λ and for simplicity we assume that the spectrum (2.6.16) is non-degenerate so that the eigenfunction u_l corresponding to the eigenvalue E_l is equal to v_l for $\lambda = 0$, i.e. the expansion coefficients $a_m(l)$ of u_l are

$$a_m(l) = \delta_{lm} + \lambda a_m{}^{(1)} + \lambda^2 a_m{}^{(2)} + \dots . \tag{2.6.23}$$

If (2.6.22) and (2.6.23) are introduced into (2.6.19), we find the following system of equations:

$$\begin{aligned}[][E_l{}^{(0)} - E_n{}^{(0)} + \lambda E_l{}^{(1)} + \lambda^2 E_l{}^{(2)} + \dots] (\delta_{nl} + \lambda a_n{}^{(1)} + \dots) \\ = \lambda \sum_m \omega_{nm} (\delta_{ml} + \lambda a_m{}^{(1)} + \dots).\end{aligned} \tag{2.6.24}$$

Since (2.6.24) is an identity in λ, we find for $l = n$

$$\begin{aligned} E_l{}^{(1)} &= \omega_{ll} \\ E_l{}^{(2)} + E_l{}^{(1)} a_l{}^{(1)} &= \sum_m \omega_{lm} a_m{}^{(1)}. \end{aligned} \tag{2.6.25}$$

For $l \neq n$ we obtain correspondingly

$$a_n{}^{(1)}(E_l{}^{(0)} - E_n{}^{(0)}) = \omega_{nl}. \tag{2.6.26}$$

By means of the equations (2.6.25)–(2.6.26) one may determine all the quantities $E_l{}^{(i)}$ and $a_m{}^{(i)}$ except $a_l{}^{(i)}$. The latter expansion coefficients must be determined from the normalization condition for u_l. Hence to first order in λ the energy E_l is given by

$$\begin{aligned} E_l &= E_l{}^{(0)} + \lambda \omega_{ll} \\ &= E_l{}^{(0)} + \Omega_{ll} \\ &= \langle E \rangle_l, \end{aligned} \tag{2.6.27}$$

which is the mean value of H in the state u_l. The expectation value $\langle E \rangle$ thus gives an estimate of the energy which is correct to first order in agreement with the Ritz variation principle.

Moreover, one finds from (2.6.23) and (2.6.26) that the wave function u_l to the first order in λ is given by

$$u_l(\mathbf{r}) = v_l(\mathbf{r}) + \sum_{n \neq l} \frac{\Omega_{nl}}{E_l{}^{(0)} - E_n{}^{(0)}} v_n(\mathbf{r}). \tag{2.6.28}$$

The wave function (2.6.28) is normalized to first order and there is no necessity of introducing the term $\lambda a_l{}^{(1)} v_l$.

The higher order corrections to these results may be found by introducing (2.6.27) and (2.6.28) in (2.6.25). For the energy to the second order, we find that

$$E_l = E_l^{(0)} + \Omega_{ll} + \sum_{m \neq l} \frac{\Omega_{lm}\Omega_{ml}}{E_l^{(0)} - E_m^{(0)}}. \tag{2.6.29}$$

Problems

2.6.1 Consider the perturbation theory for the energy E_1 of the ground state of a system. Show that the second order expression (2.6.29) leads to a lower energy than the first order result (2.6.27) if the operator $\hat{\Omega}$ has non-vanishing matrix elements Ω_{1l}.

2.6.2 Consider the one-dimensional harmonic oscillator, i.e.

$$\hat{H}_0 = -\frac{\hbar^2}{2m}\frac{\mathrm{d}^2}{\mathrm{d}x^2} + \tfrac{1}{2}m\omega^2 x^2$$

with a perturbation

$$\Omega = \lambda x.$$

Show that the second order perturbation calculation leads to

$$E_l = (l + \tfrac{1}{2})\hbar\omega - \frac{\lambda^2}{2m\omega^2}$$

[utilize the result of problem 2.4.3] and verify that this agrees with an exact calculation.

3

Time Dependent Behaviour of Particles

3.1 THE MOTION OF A WAVE PACKET. EHRENFEST'S THEOREM

In the following sections we will discuss the passage of particles through various potentials. Since we will be interested especially in typical quantum mechanical effects, we will first try to sort out those cases where quantum mechanics automatically yields the same answers as classical mechanics. To this end we examine how the mean values of the coordinates change in time, because in this way we get information about the motion of a narrow wave packet.

In 1.6 we calculated the time derivative of $\langle E \rangle$; the same method can be used also to calculate time derivatives of the mean values of other quantities. We will thus consider a physical quantity q, whose corresponding operator we will call $\hat{q}$. We find

$$\frac{\mathrm{d}}{\mathrm{d}t}\langle q\rangle = \frac{\mathrm{d}}{\mathrm{d}t}\int \psi^*\hat{q}\psi\,\mathrm{d}\tau = \int\left[\left(\frac{\partial\psi}{\partial t}\right)^*\hat{q}\psi + \psi^*\hat{q}\,\frac{\partial\psi}{\partial t}\right]\mathrm{d}\tau, \quad (3.1.1)$$

provided $\hat{q}$ does not depend explicitly on time. We then use the Schrödinger equation (1.6.14) and find

$$\frac{\mathrm{d}}{\mathrm{d}t}\langle q\rangle = -\frac{1}{\mathrm{i}\hbar}\int (\hat{H}\psi)^*\hat{q}\psi\,\mathrm{d}\tau + \frac{1}{\mathrm{i}\hbar}\int \psi^*\hat{q}\hat{H}\psi\,\mathrm{d}\tau, \quad (3.1.2)$$

where $\hat{H}$ in the first integral only acts on ψ^*.

Utilizing the fact that $\hat{H}$ is a Hermitian operator (1.6.20) we find

$$\frac{\mathrm{d}}{\mathrm{d}t}\langle q\rangle = \frac{1}{\mathrm{i}\hbar}\int \psi^*(\hat{q}\hat{H} - \hat{H}\hat{q})\psi\,\mathrm{d}\tau. \quad (3.1.3)$$

We can write this result in the form

$$\frac{\mathrm{d}}{\mathrm{d}t}\langle q\rangle = \frac{1}{\mathrm{i}\hbar}\langle[\hat{q},\hat{H}]\rangle, \tag{3.1.4}$$

where we have introduced the so-called commutator of $\hat{q}$ and $\hat{H}$ through the definition

$$[\hat{q},\hat{H}] = \hat{q}\hat{H}-\hat{H}\hat{q}. \tag{3.1.5}$$

Let us first consider the time derivative of $\langle x\rangle$. We find directly from (3.1.4) that

$$\frac{\mathrm{d}}{\mathrm{d}t}\langle x\rangle = \frac{1}{\mathrm{i}\hbar}\langle x\hat{H}-\hat{H}x\rangle. \tag{3.1.6}$$

When we compute the commutator of x and $\hat{H}$, the potential energy gives no contribution since evidently $x\Phi(\mathbf{r}) = \Phi(\mathbf{r})x$; x and Φ are commutable operators. In the Laplacian operator Δ we can similarly disregard the terms $\partial^2/\partial y^2$, $\partial^2/\partial z^2$, as they evidently are interchangeable with x, and we therefore get

$$\begin{aligned}(x\hat{H}-\hat{H}x)\psi &= -\frac{\hbar^2}{2m}\left(x\frac{\partial^2}{\partial x^2}-\frac{\partial^2}{\partial x^2}x\right)\psi \\ &= -\frac{\hbar^2}{2m}\left(x\frac{\partial^2\psi}{\partial x^2}-2\frac{\partial\psi}{\partial x}-x\frac{\partial^2\psi}{\partial x^2}\right) = \frac{\hbar^2}{m}\frac{\partial}{\partial x}\psi.\end{aligned} \tag{3.1.7}$$

We thus find, according to (1.6.6),

$$\frac{1}{\mathrm{i}\hbar}[x,\hat{H}] = \frac{1}{m}\hat{p}_x,$$

which by insertion in (3.1.4) gives

$$\frac{\mathrm{d}}{\mathrm{d}t}\langle x\rangle = \frac{1}{m}\langle p_x\rangle. \tag{3.1.8}$$

Now, we can easily calculate the second time derivative of $\langle x\rangle$:

$$\frac{\mathrm{d}^2}{\mathrm{d}t^2}\langle x\rangle = \frac{1}{m}\frac{\mathrm{d}}{\mathrm{d}t}\langle p_x\rangle = \frac{1}{\mathrm{i}m\hbar}\langle[\hat{p}_x,\hat{H}]\rangle. \tag{3.1.9}$$

Since $\hat{p}_x = -i\hbar(\partial/\partial x)$ is interchangeable with the operator Δ, one gets

$$(\hat{p}_x \hat{H} - \hat{H}\hat{p}_x)\psi = -i\hbar\left(\frac{\partial}{\partial x}\Phi(\mathbf{r}) - \Phi(\mathbf{r})\frac{\partial}{\partial x}\right)\psi = -i\hbar\frac{\partial\Phi}{\partial x}\psi, \tag{3.1.10}$$

which by insertion in (3.1.9) gives

$$\frac{d^2}{dt^2}\langle x\rangle = \frac{1}{m}\left\langle -\frac{\partial\Phi}{\partial x}\right\rangle = \frac{1}{m}\langle K_x(\mathbf{r})\rangle, \tag{3.1.11}$$

where K_x is the x-component of the force, $\mathbf{K} = -\text{grad}\,\Phi$. The equations (3.1.8) and (3.1.11) and the analogous equations for the other coordinates allow a precise formulation of the correspondence between quantum mechanics and classical mechanics, since they show that if $\langle\mathbf{K}(\mathbf{r})\rangle = \mathbf{K}(\langle\mathbf{r}\rangle)$, the mean values $\langle\mathbf{r}\rangle$, $\langle\mathbf{p}\rangle$ and $\langle\mathbf{K}\rangle$ of quantum mechanics satisfy the same equations as $\mathbf{r}$, $\mathbf{p}$ and $\mathbf{K}$ in classical mechanics. This result was first deduced by P. Ehrenfest.

By using (3.1.8) on a wave packet, whose position can be represented by the mean value of the coordinates, we find that the velocity of the wave packet in the direction of the x-axis is $\langle p_x\rangle/m$. From (3.1.11) we find the acceleration of the wave packet. If the dimension of the wave packet is small in comparison to the distances over which the force changes perceptibly, then the force can be considered constant throughout the wave packet, and it follows that $\langle\mathbf{K}(\mathbf{r})\rangle = \mathbf{K}(\langle\mathbf{r}\rangle)$. In this case (3.1.11) will become identical with the classical equations of motion for a particle with position vector $\langle\mathbf{r}\rangle$, and the wave packet will be accelerated according to classical mechanics. According to the uncertainty relations, however, the wave packet cannot be confined to a region with dimensions less than $\delta x \approx h/\delta p_x$, $\delta y \approx h/\delta p_y$, $\delta z \approx h/\delta p_z$. If the state is to have a well-defined momentum, then the 'fuzziness' in the momentum components must be small in comparison with the magnitude, $\approx p$, of the momentum. Consequently, the extension of the wave packet must be large in comparison with the de Broglie wavelength $\lambda = h/p$, and the wave packet will only move as a classical particle if the variation of the force over a wavelength is negligible. It is important to notice that this criterion does not restrict the validity of classical mechanics only to macroscopic particles. We know that the motion of electrons in spectographs and accelerators can be described accurately on the basis of classical mechanics, and this is seen to be in accordance with the above criterion considering that the wavelength of the electrons is usually less than 10^{-8} cm, while the electromagnetic fields in an apparatus are defined by a macroscopic set-up.

We have here given only a qualitative description of the motion of a wave packet, since we have considered only the changes of the mean value $\langle x \rangle$ and have disregarded the change in time of the shape of the wave packet. A more complete investigation should include also a discussion of this latter question since a broadening of an originally narrow wave packet may imply that the acceleration (3.1.8) gradually deviates from the classical one.

The question of the smearing out of wave packets in the course of time can be treated by methods similar to those used above. One can calculate the time derivatives of $\langle x^2 \rangle$, and thereby determine the changes in fluctuation $\sigma_x^2 = \langle x^2 \rangle - \langle x \rangle^2$, when the time derivatives of $\langle x \rangle$ are known. We shall here not attempt such a general investigation which leads to somewhat laborious calculations. In the following sections we will see examples of violent changes in the shape of a wave packet: a rapidly changing potential may cause the packet to break up into two parts. It should, however, be emphasized that even when the force is constant over a wave packet, it may be smeared out in the long run. As an illustration we will discuss only the smearing out for a free particle. One finds that $\mathrm{d}\sigma_x^2/\mathrm{d}t$ depends in a complicated manner on the state, while the derivative of second order is given by the simple expression $\mathrm{d}^2\sigma_x^2/\mathrm{d}t^2 = 2\sigma_p^2/m^2$ [problem 3.1.4]. All derivatives of order higher than 2 are zero because σ_p^2 is constant for a free particle [problem 3.1.1] and one then finds

$$\sigma_x^2(t) = \tfrac{1}{2}\frac{\mathrm{d}^2\sigma_x^2}{\mathrm{d}t^2}t^2 = \frac{1}{m^2}\sigma_p^2(0)t^2,$$

where we have assumed that $\mathrm{d}\sigma_x^2/\mathrm{d}t = 0$ for $t = 0$.

As an estimate of the variation of σ_x in time we can therefore use

$$\sigma_x(t) \approx \left(\sigma_x^2(0) + \frac{\sigma_p^2(0)}{m^2}t^2\right)^{\frac{1}{2}}. \tag{3.1.12}$$

This expression has a very obvious meaning since σ_p/m is a measure of the fluctuation in velocity and consequently $\sigma_p t/m$ is a measure of the fluctuation in the distance travelled. The uncertainty in momentum at the time $t = 0$ gives rise to a fluctuation in the distance travelled which is proportional to time, and according to (3.1.12) this must be added to the original fluctuation in the coordinates in order to yield the width of the wave packet. This purely kinematic smearing of the wave packet will appear even when the forces do not change the shape of the wave packet. In the special case of a harmonic oscillator the forces prevent the smearing over a long time, each state in this case having a strictly periodic time-variation.

It should be noted that also according to classical mechanics the uncertainty (3.1.12) would result for the x-coordinate of a particle which at an initial time t has an uncertainty $\sigma_x(0)$ in the determination of the position and $\sigma_p(0)/m$ in the determation of the velocity. Quantum mechanics only play a rôle in the interpretation of (3.1.12) when one takes into consideration that σ_p cannot be less than $\hbar/(2\sigma_x)$. From (3.1.12) one then finds

$$\sigma_x(t) \geqq \left(\sigma_x{}^2(0)+\frac{\hbar^2 t^2}{4m^2\sigma_x{}^2(0)}\right)^{\frac{1}{2}} = \sigma_x(0)\left(1+\frac{\hbar^2 t^2}{4m^2\sigma_x{}^4(0)}\right)^{\frac{1}{2}},$$

which shows that for a chosen $\sigma_x(0)$ the smearing of the wave packet is only negligible during a time which is small in comparison to $m\sigma_x{}^2(0)/\hbar$ [see problem 1.2.1]. For a fixed time there exists an optimal starting fluctuation which minimizes $\sigma_x(t)$.

Problems

3.1.1 Show, in the case of a free particle, that the momentum distribution is constant in time. (Calculate, for example, the time derivatives of arbitrary powers of the momentum components.)

3.1.2 Use the theorem of Ehrenfest to show that for a one-dimensional harmonic oscillator in an arbitrary state, $\langle x\rangle$ varies harmonically in time with the classical frequency [compare with problem 2.5.4].

3.1.3 Show, for a particle under the influence of a constant force in the entire space, that $\langle \mathbf{r}\rangle$ changes in time exactly like the position of a classical particle.

3.1.4 Show that the following equations are fulfilled for a free particle:

$$\frac{\mathrm{d}}{\mathrm{d}t}\sigma_x{}^2 = \frac{1}{m}\int \psi^*(x\hat{p}_x+\hat{p}_x x)\psi \mathrm{d}\tau - \frac{1}{m}\langle x\rangle\,\langle p_x\rangle,$$

$$\frac{\mathrm{d}^2}{\mathrm{d}t^2}\sigma_x{}^2 = \frac{2}{m^2}\sigma_p{}^2.$$

Show that for a wave function of the form $\exp(ikx)F(x)$, where F is real one has

$$\frac{\mathrm{d}}{\mathrm{d}t}\sigma_x{}^2 = 0.$$

3.1.5 The following formula applies in first approximation for a narrow one-dimensional wave packet:

$$\frac{d^2}{dt^2}\langle x\rangle = \frac{1}{m}K(\langle x\rangle).$$

Show that in the next approximation one must add a term $K''(\langle x\rangle)\sigma_x{}^2/2m$.

3.2 ONE-DIMENSIONAL SCATTERING PROBLEMS. PROBABILITY CURRENT

Scattering experiments play an important rôle in the exploration of the structure of atoms and nuclei. Through measurements of cross-sections for various deflection angles one is able to make conclusions about the interactions which the projectiles are exposed to during their passage through the target. Rutherford's discovery of the nucleus stands as a classical example of the significance of this method. In atomic scattering problems one often has to deal with potentials which vary rapidly within a wavelength of the projectiles, and one must therefore usually treat the scattering quantum mechanically. In this section we will only illustrate the principles of such a quantum mechanical treatment by considering the simplest one-dimensional potential scattering.

The potential along the x-axis may, for instance, have the shape illustrated in Fig. 3.1. One notes that the particle is free for $x \to \pm\infty$ as $\Phi(x)$ tends asymptotically towards constant values 0 and Φ_0. The points a and b which define the region where forces appear divide the x-axis into three regions I, II and III. The scattering

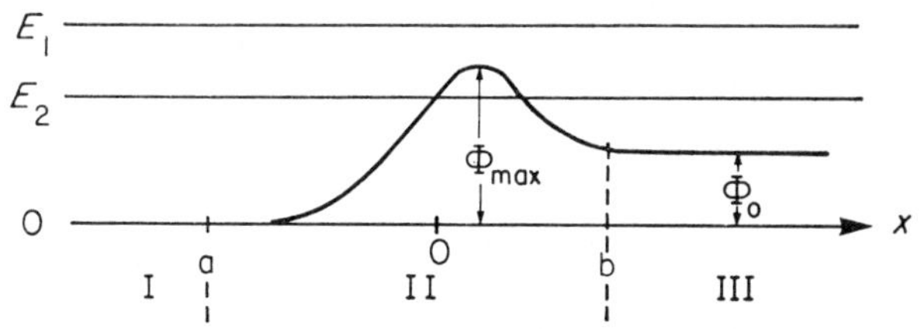

Fig. 3.1

problem may be formulated as follows. A particle moves from region I towards II. In which manner will the particle continue its motion over a long time period? From the point of view of classical mechanics, the answer is completely trivial. If the total energy, E, of the particle is larger than Φ_{max}, the particle will continue into III and move on along the x-axis with a velocity corresponding to the kinetic energy $E-\Phi_0$. If, however, $E<\Phi_{max}$, the particle will be reflected from the potential barrier and return to region I.

In quantum mechanics the situation is more complicated. As we shall see in the following sections, there is here a finite probability that the particle is reflected even though $E>\Phi_{max}$, and there will be a finite probability that it is transmitted through II, although $E<\Phi_{max}$. In this section we will discuss the general method of determining the probability of reflexion or transmission.

We base our investigation on the stationary Schrödinger equation, which in a one-dimensional problem takes the form (2.1.17). If we confine ourselves to the case $E>\Phi_0$, the equation has solutions of the type (2.3.4) in the regions I and III. In the present context it is however more convenient to use the complex solution, which can be written as

$$u_1 = A\exp(ik_1x)+B\exp(-ik_1x), \quad k_1 = \left(\frac{2mE}{\hbar^2}\right)^{\frac{1}{2}}, \tag{3.2.1}$$

$$u_2 = C\exp(ik_2x)+D\exp(-ik_2x), \quad k_2 = \left(\frac{2m(E-\Phi_0)}{\hbar^2}\right)^{\frac{1}{2}}. \tag{3.2.2}$$

In the region II the solution depends on the special form of the potential; like the asymptotic solutions it will in any case contain two integration constants. A complete solution to the Schrödinger equation is found by matching the three solutions through the continuity conditions at the points a and b. This places four restraints on the six constants of integration. Consequently, for a given value of E, one may choose the ratio of two of the constants in advance, whereafter $u(x)$ is fixed on the entire x-axis apart from a constant normalization factor.

In order to describe a particle moving towards the potential barrier we will build up, by superposition, a wave packet of the stationary solutions. By following the development in time of this state we learn what happens to the particle. If the superposition only contains states in an interval, δE, around a certain value, E_0, one is able to consider the energy in the time-dependent state as well-defined and equal to E_0 with any desired degree of accuracy only if δE is chosen sufficiently small.

In the solutions (3.2.1) and (3.2.2) all terms are of the form $\exp[\mathrm{i}k(E)x]$, where $k(E) = [2m(E-\Phi)/\hbar^2]^{\frac{1}{2}}$ with Φ constant. We will therefore start by examining superpositions of functions of this type. An approximate well-defined energy is ensured by letting the waves appear in the superposition with a weight factor $g(E-E_0)$, which is different from zero only for $|E-E_0|<\delta E/2$. According to (2.5.18) the resulting time-dependent wave function is given by

$$\psi_\pm(x, t) = \int g(E-E_0) \exp[\pm \mathrm{i}k(E)x - \mathrm{i}Et/\hbar]\, \mathrm{d}E. \qquad (3.2.3)$$

When δE is sufficiently small, the function $k(E)$ can be substituted by the two first terms in a series expansion around E_0. We may thus write

$$k(E) = k(E_0)+\frac{\mathrm{d}k}{\mathrm{d}E}(E-E_0) = k(E_0)+\frac{m}{\hbar^2 k(E_0)}(E-E_0). \qquad (3.2.4)$$

The quantity $\hbar k(E_0)/m = [2(E_0-\Phi)/m]^{\frac{1}{2}}$ is equal to the numerical value of the classical velocity. By using (3.2.4) and the abbreviation $v = \hbar k(E_0)/m$, (3.2.3) can be written in the form

$$\psi_\pm(x, t) = \exp[\pm \mathrm{i}k(E_0)x - \mathrm{i}E_0 t/\hbar] \int g(E-E_0) \exp\left[\frac{\mathrm{i}}{\hbar}\left(\pm\frac{x}{v}-t\right)(E-E_0)\right] \mathrm{d}(E-E_0) \qquad (3.2.5)$$

where the integral in (3.2.5) depends only on x and t through the quantity $\pm(x/v)-t$. The wave function (3.2.5) can thus be written

$$\psi_\pm(x, t) = \exp[\pm \mathrm{i}k(E_0)x - \mathrm{i}E_0 t/\hbar]\, F\left(\pm\frac{x}{v}-t\right), \qquad (3.2.6)$$

where the accurate form of $F(\xi)$ depends on the choice of the weight function $g(E-E_0)$. For the sake of convenience we choose g real and symmetric around E_0 whereby $F(\xi)$ becomes real and even in ξ. The representation (3.2.5) of $F(\xi)$ as a Fourier integral also shows that $F(\xi)$ can vanish outside an interval of the order of magnitude $\hbar/\delta E$, since $g(E-E_0)$ differs from zero inside an interval δE. The wave function (3.2.6) thus describes a wave packet of extent $\approx v\hbar/\delta E$ with its centre at $x = \pm vt$. Apparently the wave packet propagates with the velocity v in the positive or negative direction according to whether the state is formed by superposition of functions $\exp(+ikx)$ or $\exp(-ikx)$.

We are now able to describe the behaviour of a particle which comes from I and moves in towards the potential barrier, by superposition of wave-functions of type (3.2. 1–2), putting $D = 0$.

The stationary solutions corresponding to the energy E are then

$$
\begin{aligned}
u_E(x) &= A \exp[\mathrm{i}k_1(E)x] + B(E)\exp[-\mathrm{i}k_1(E)x] \quad &\text{for } x<a,\\
u_E(x) &= C(E)\exp[\mathrm{i}k_2(E)x] \quad &\text{for } x>b,
\end{aligned}
\tag{3.2.7}
$$

where B and C, for fixed A, only depend on the energy. We now make a superposition of these stationary solutions corresponding to energies in a small interval δE surrounding E_0, and the wave function is then found to be

$$
\psi(x,t) = \int g(E-E_0)u_E(x)\exp\left(-\frac{\mathrm{i}}{\hbar}Et\right)\mathrm{d}E. \tag{3.2.8}
$$

When inserting (3.2.7) in (3.2.8) we are able, for δE sufficiently small, to consider B and C as constants [$B(E_0)$ and $C(E_0)$], which may be moved outside the sign of integration. By using the same manipulations which led us from (3.2.3) to (3.2.6) we find that (3.2.8) can be approximated by

$$
\begin{aligned}
\psi(x,t) = {} & A\exp[\mathrm{i}k_1(E_0)x - \mathrm{i}E_0t/\hbar]\, F\left(\frac{x}{v_1}-t\right)\\
&+ B\exp[-\mathrm{i}k_1(E_0)x - \mathrm{i}E_0t/\hbar]\, F\left(\frac{x}{v_1}+t\right), \\
&\qquad \text{for } x<a;
\end{aligned}
\tag{3.2.9}
$$

$$
\psi(x,t) = C\exp[\mathrm{i}k_2(E_0)x - \mathrm{i}E_0t/\hbar]\, F\left(\frac{x}{v_2}-t\right) \qquad \text{for } x>b, \tag{3.2.10}
$$

where we have made use of $F(\xi) = F(-\xi)$.

The wave function (3.2.9), (3.2.10), represents three wave packets with coefficients A, B and C. It follows from the properties of the

F-function that in a remote past, $t = -t_0(t_0 \gg h/\delta E)$, the A-packet is situated at $x = -v_1 t_0$, completely within the region I. The B-packet lies at $x = v_1 t_0$, where (3.2.9) does not apply, and the C-packet lies at $x = -v_2 t_0$, where (3.2.10) does not apply. Consequently, at large negative times, (3.2.9) and (3.2.10) only represent a single wave packet, which in the region I travels with velocity v towards the right. The corresponding probability density is

$$|\psi|^2 = |A|^2 F^2\left(\frac{x}{v_1} - t\right), \quad \text{if } t \ll -h/\delta E. \tag{3.2.11}$$

In the course of time the A-packet, describing the incoming particle, disappears from I, and at a time $t \gg +h/\delta E$, (3.2.9) and (3.2.10) represent two wave packets, a transmitted one and a reflected one. The probability density is then given by

$$|\psi|^2 = |B|^2 F^2\left(\frac{x}{v_1} + t\right) + |C|^2 F^2\left(\frac{x}{v_2} - t\right), \quad \text{if } t \gg +h/\delta E. \tag{3.2.12}$$

The two situations are indicated in Fig. 3.2.

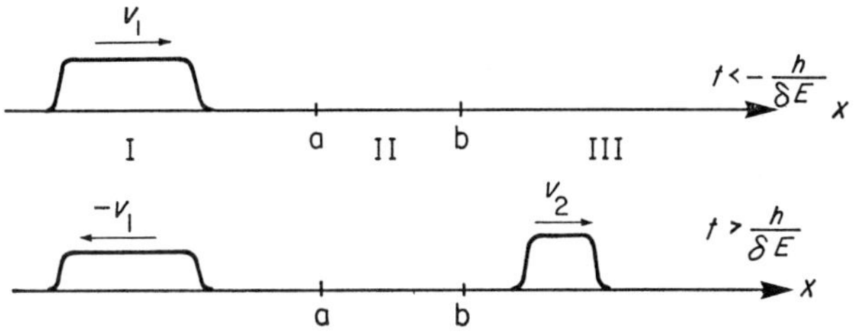

Fig. 3.2

One may choose the weight function used in the superposition in such a way that the incoming wave packet is normalized. According to (3.2.11) one then demands that

$$|A|^2 \int_{-\infty}^{\infty} F^2\left(\frac{x}{v_1} - t\right) \mathrm{d}x = 1$$

or

$$\int F^2(\xi)\mathrm{d}\xi = \frac{1}{|A|^2 v_1}. \tag{3.2.13}$$

The probability of finding the particle at late times in III (the transmission probability, T) can then be found from (3.2.12) as

$$T = |C|^2 \int F^2\left(\frac{x}{v_2} - t\right) \mathrm{d}x = |C|^2 v_2 \int F^2(\xi) \mathrm{d}\xi,$$

which, using (3.2.13), becomes

$$T = \frac{|C|^2 v_2}{|A|^2 v_1} = \frac{|C|^2 k_2(E_0)}{|A|^2 k_1(E_0)}. \tag{3.2.14}$$

In a corresponding manner the reflexion probability (or reflexion coefficient) is found from (3.2.12) and (3.2.13):

$$R = \frac{|B|^2 v_1}{|A|^2 v_1} = \frac{|B|^2}{|A|^2}. \tag{3.2.15}$$

It follows from the time independence of the normalization integral that A, B and C must be related in such a way that $R+T = 1$.

We note that the resulting expressions for T and R do not depend on the explicit form of the incoming wave packet. The expressions (3.2.14) and (3.2.15) only contain quantities which refer to the stationary solution for the energy E_0. Thus the results may also be deduced directly from the stationary solution without using the somewhat tedious but more picturesque method of superposition which we used above.

One gets the stationary wave function corresponding to the energy E_0 from (3.2.7) by multiplication by the time factor $\exp(-\mathrm{i}E_0 t/\hbar)$. This stationary wave function is improper and corresponds to a probability density which is constant in time. It may, however, be interpreted as a limiting case of (3.2.8) corresponding to $\delta E \to 0$, and hence one may imagine that the stationary state results in the limit from the wave packets mentioned above when these are made infinitely long. One sees that also in the stationary state there must exist probability currents with the same velocities as for the long wave packets. This is not inconsistent with the constancy of the probability density, provided the probability current is stationary.

According to the argument above a wave function $A \exp(ikx)$ can be associated with a probability current, j, which is proportional to the probability density and the classical velocity (the group velocity) $v = \hbar k/m$, so that one can put $j = |A|^2 v = |A|^2 \hbar k/m$. The stationary solution (3.2.7) thus represents, for $x<a$, and incoming

current ($j_i = |A|^2 v_1$) plus a reflected current ($j_r = |B|^2 v_1$), and for $x > b$ a transmitted current ($j_{tr} = |C|^2 v_2$). If we maintain the picture of the scattering process as a time-dependent phenomenon, this stationary situation may be interpreted as an approximate description of the situation during a time interval $-h/\delta E \ll t \ll h/\delta E$, lying between the two situations pictured in Fig. 3.2.

In the stationary description the probabilities T and R must be defined as those fractions of the incoming current which are transmitted or reflected respectively. One thus finds

$$T = \frac{|j_{tr}|}{|j_i|}, \quad R = \frac{|j_r|}{|j_i|}, \tag{3.2.16}$$

which, by insertion of the current densities, becomes identical to (3.2.14) and (3.2.15).

We have now introduced the concept of probability current in an informal manner. A more rigorous definition can be based on (1.3.7), which expresses the change in time of the total probability inside an arbitrary region, Ω, confined within a surface F:

$$\frac{\mathrm{d}}{\mathrm{d}t}\int_{\Omega} |\psi|^2 \mathrm{d}\tau = \frac{\hbar}{2im}\int_{F} (\psi \operatorname{grad} \psi^* - \psi^* \operatorname{grad} \psi)\mathrm{d}\mathbf{f}. \tag{3.2.17}$$

Introducing the abbreviations $P(\mathbf{r}, t) = |\psi|^2$ and

$$\mathbf{j}(\mathbf{r}, t) = \frac{\hbar}{2im}(\psi^* \operatorname{grad} \psi - \psi \operatorname{grad} \psi^*), \tag{3.2.18}$$

(3.2.17) can be written

$$\frac{\mathrm{d}}{\mathrm{d}t}\int_{\Omega} P \cdot \mathrm{d}\tau = -\int_{F} \mathbf{j} \cdot \mathrm{d}\mathbf{f}. \tag{3.2.19}$$

One sees that $\mathbf{j}$ must be identified with the probability current density, and (3.2.19) then expresses that the increase per second of probability in the domain Ω is equal to the probability which during the same time has passed through the boundary surface F. The relation (3.2.19) gives a precise meaning—also for improper wave functions—to the interpretation of $|\psi|^2$ as a measure of the probability density inside limited regions. Using Gauss's theorem on the right hand side of equation

(3.2.19) one is able to deduce the differential continuity equation

$$\frac{\partial P}{\partial t}+\operatorname{div} \mathbf{j} = 0, \tag{3.2.20}$$

which is analogous to the continuity equation in electrodynamics. For a stationary state $P = |\psi|^2$ is independent of time and one therefore finds, according to (3.2.20), that div $\mathbf{j} = 0$. In the one-dimensional scattering problem considered above this means that the current must be constant on the entire x-axis, and consequently $j_i+j_r = j_{tr}$. In terms of the magnitude of the currents one finds $|j_i| = |j_{tr}|+|j_r|$, which inserted in (3.2.16) gives $T+R = 1$.

Problems

3.2.1 Find, from the definition (3.2.18), the current component j_x for a wave function A exp (ikx), and find $\mathbf{j}$ for a wave function A exp $(ikx)/r$, where $r^2 = x^2+y^2+z^2$. In the latter case find the total outgoing current through a spherical surface with radius R and centre (0, 0, 0). Verify that any real wave function corresponds to $\mathbf{j} = 0$ throughout space.

3.2.2 A particle moves freely on the negative part of the x-axis. At $x = 0$ there exists an infinitely high potential wall. Find the stationary wave functions belonging to the energy E. Show that when the particle moves towards the wall it is reflected with the probability $R = 1$ for all values of E.

3.2.3 Show that the result found in the last problem is generally valid for the potential in Fig. 3.1 when the energy of the incoming particle is less than Φ_0.

3.2.4 In the potential in Fig. 3.1, a particle with energy $E>\Phi_0$ is considered. When the particle arrives from the left the reflexion and transmission coefficients are called R_l and T_l. The corresponding quantities for arrivals from the right are R_r and T_r. Show, that in general one has $R_l = R_r$ and $T_l = T_r$. (Use the fact that the stationary wave functions in the two cases are solutions of the stationary Schrödinger equation for the same energy.)

3.3 POTENTIAL WALL AND RECTANGULAR BARRIER

As an illustration of the general discussion in the previous section we will first consider the passage of a particle over a potential wall as shown in Fig. 3.3. The problem may be regarded as an idealization of the passage of an electron perpendicular to the surface of a metal.

We will only treat particles which arrive from the left and we will disregard the case $E<\Phi_0$, which leads to 100% reflexion [problem

3.2.3]. For $E > \Phi_0$ the solution of the stationary Schrödinger equation, in the two regions I and II, are found to be

$$u_{\mathrm{I}} = A \exp(\mathrm{i}k_1 x) + B \exp(-\mathrm{i}k_1 x), \quad k_1 = \left(\frac{2mE}{\hbar^2}\right)^{\frac{1}{2}} \tag{3.3.1}$$

$$u_{\mathrm{II}} = C \exp(\mathrm{i}k_2 x), \; k_2 = \left(\frac{2m(E-\Phi_0)}{\hbar^2}\right)^{\frac{1}{2}} \tag{3.3.2}$$

when we describe particles arriving from the left.

Application of the continuity conditions at the point $x = 0$ gives, for (3.3.1) and (3.3.2)

$$A + B = C;$$

$$\mathrm{i}k_1(A-B) = \mathrm{i}k_2 C,$$

leading to

$$\frac{B}{A} = \frac{(1-k_2/k_1)}{(1+k_2/k_1)}, \qquad \frac{C}{A} = \frac{2}{1+k_2/k_1}. \tag{3.3.3}$$

For the transmission and reflexion probabilities one finds then from (3.3.3), (3.2.14) and (3.2.15)

$$T = \frac{k_2}{k_1}\frac{|C|^2}{|A|^2} = \frac{4k_1k_2}{(k_1+k_2)^2}, \quad R = \left|\frac{B}{A}\right|^2 = \frac{(k_1-k_2)^2}{(k_1+k_2)^2}, \tag{3.3.4}$$

where k_1 and k_2 are given by (3.3.1) and (3.3.2).

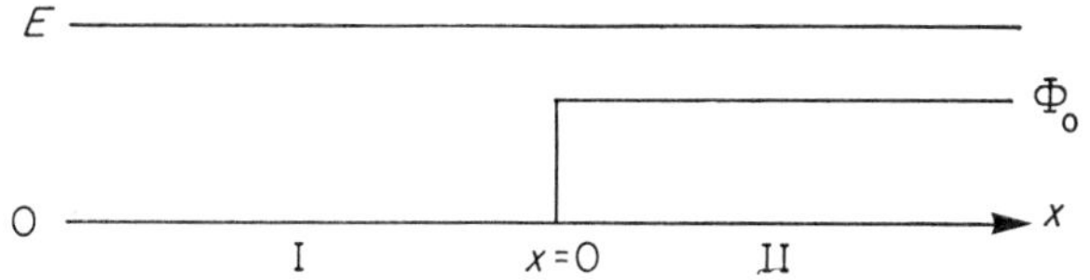

Fig. 3.3

It is easily verified directly from (3.3.4) that $T+R = 1$, as was to be expected. It is also seen that T and R are symmetric in k_1 and k_2. This means that one would find the same reflexion probability if the potentials in I and II were interchanged [cf. problem 3.2.4].

Fig. 3.4 shows how the probabilities (3.3.4) vary as functions of

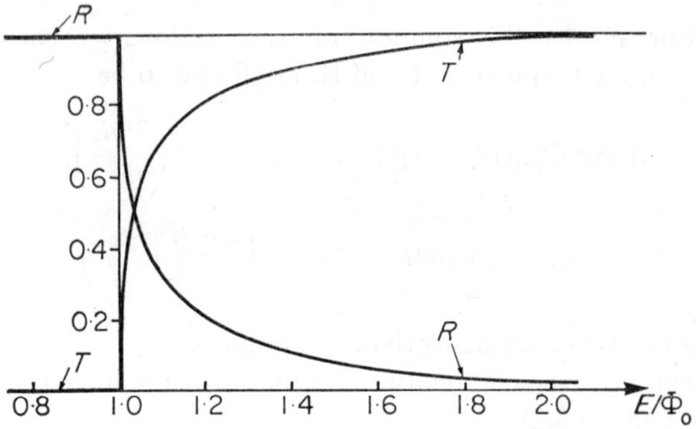

Fig. 3.4

E/Φ_0. If either $E<\Phi_0$ or $E\gg\Phi_0$, the results are seen to agree with classical expectations.

As another example of a one-dimensional scattering problem, we will now consider a particle moving towards an idealized potential barrier of height Φ_0 and width b (see Fig. 3.5).

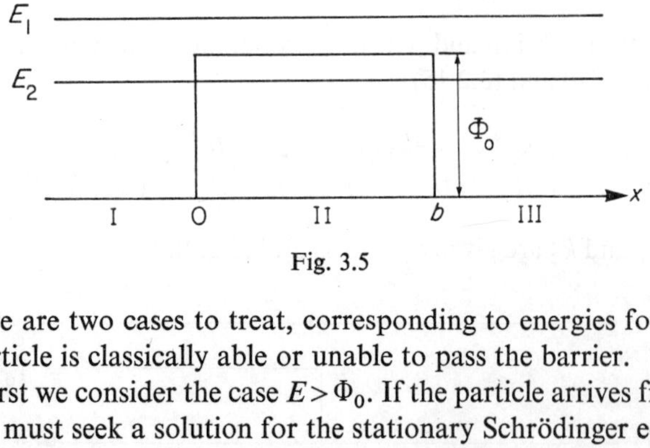

Fig. 3.5

There are two cases to treat, corresponding to energies for which the particle is classically able or unable to pass the barrier.

At first we consider the case $E>\Phi_0$. If the particle arrives from the left we must seek a solution for the stationary Schrödinger equation which, in the region III, only describes particles moving towards the right. The solution in the three regions is thus

$$\begin{aligned} u_{\mathrm{I}} &= \exp(ikx)+B\exp(-ikx) & k &= \left(\frac{2mE}{\hbar^2}\right)^{\frac{1}{2}} \\ u_{\mathrm{II}} &= F\exp(ik'x)+G\exp(-ik'x) & k' &= \left(\frac{2m(E-\Phi_0)}{\hbar^2}\right)^{\frac{1}{2}} \\ u_{\mathrm{III}} &= C\exp(ikx). \end{aligned} \tag{3.3.5}$$

For convenience we have chosen the arbitrary factor in the solution in such a way that the incoming current is represented by exp (ikx). The reflexion coefficient then becomes $|B|^2$, and since the kinetic energy is the same in the two regions I and III, the transmission coefficient is simply given by $T = |C|^2$.

The continuity conditions for u and u' at $x = 0$ and $x = b$ yield the four equations

$$\left.\begin{aligned} 1+B &= F+G, && \text{(a)}\\ ik(1-B) &= ik'(F-G), && \text{(b)}\\ F\exp(ik'b)+G\exp(-ik'b) &= C\exp(ikb), && \text{(c)}\\ ik'[F\exp(ik'b)-G\exp(-ik'b)] &= ikC\exp(ikb). && \text{(d)} \end{aligned}\right\} \quad (3.3.6)$$

The amplitudes F and G can be found in terms of the other parameters from (a) and (b) as well as (c) and (d). When the two expressions for F and G are equated, one obtains

$$\begin{aligned} 2F &= \left(1+\frac{k}{k'}\right)+B\left(1-\frac{k}{k'}\right) = C\exp[i(k-k')b]\left(1+\frac{k}{k'}\right)\\ 2G &= \left(1-\frac{k}{k'}\right)+B\left(1+\frac{k}{k'}\right) = C\exp[i(k+k')b]\left(1-\frac{k}{k'}\right), \end{aligned} \quad (3.3.7)$$

from which one is able to find the amplitudes, B and C, for the reflected and the transmitted waves. If one is not interested in the wave function but only in the probabilities R and T, it is sufficient to determine one of the coefficients B or C from (3.3.7) and use the relation $R+T = 1$.

From (3.3.7) one finds by eliminating B

$$C = \frac{\exp(-ikb)\,4(k/k')}{\exp(-ik'b)\,(1+k/k')^2 - \exp(ik'b)\,(1-k/k')^2}, \quad (3.3.8)$$

and the absolute square gives

$$T = |C|^2 = \frac{16(k/k')^2}{(1+k/k')^4+(1-k/k')^4-2(1+k/k')^2(1-k/k')^2\cos 2k'b},$$

which may be rewritten

$$T = \left\{1+\tfrac{1}{4}\frac{(k^2-k'^2)^2}{k^2k'^2}\sin^2(k'b)\right\}^{-1}. \quad (3.3.9)$$

By the insertion of the expressions (3.3.5) for k and k' in (3.3.9) it is convenient to introduce the abbreviation

$$\kappa = \frac{m\Phi_0 b^2}{2\hbar^2}. \tag{3.3.10}$$

The transmission coefficient (3.3.9) may then be written in the form

$$T = \left\{1+\tfrac{1}{4}\frac{\sin^2[2\kappa^{\frac{1}{2}}(E/\Phi_0-1)^{\frac{1}{2}}]}{(E/\Phi_0-1)\,E/\Phi_0}\right\}^{-1} \quad \text{for } E>\Phi_0. \tag{3.3.11}$$

From (3.3.11) one finds $T\to 1/(1+\kappa)$, $E\to\Phi_0$ whereas for $E/\Phi_0\to\infty$ the transmission coefficient approaches the classical value, $T = 1$. It should be noted, though, that T does not increase monotonically with E. There will exist complete transmission for a series of discrete E-values, since T is equal to 1 when the argument $2\kappa^{\frac{1}{2}}\,(E/\Phi_0-1)^{\frac{1}{2}}$ equals an integral multiple of π (Fig. 3.6). From (3.3.9) one sees that complete transmission occurs when the width of the barrier, b, is equal to an integral number of half wave lengths ($\lambda' = 2\pi k'$). This resonance phenomenon, which is called selective transmission is a typical quantum mechanical effect. The analogous phenomenon for three-dimensional scattering gives rise to the so-called Ramsauer effect, which occurs when electrons with very low energy (~1eV) pass through atoms of the inert gases without being appreciably affected by the strong potential in the atom.*

On the basis of the above results it is easy to treat the transmission problem for $E<\Phi_0$. For $E<\Phi_0$ one obtains, instead of the oscillating functions in the region II, the exponential functions $\exp(\pm\beta x)$, $\{\beta = [2m(\Phi_0-E)/\hbar^2]^{\frac{1}{2}}\}$, but apart from this there are no changes Consequently we can employ the previous results, replacing ik' by β. Thus $\sin(k'b)$ changes into $-i\sinh(\beta b)$ and from (3.3.9) the transmission coefficient is then found to be

$$T = \left[1+\tfrac{1}{4}\frac{(k^2+\beta^2)^2}{k^2\beta^2}\sinh^2(\beta b)\right]^{-1}.$$

Inserting the expressions for k and β yields

$$T = \left\{1+\tfrac{1}{4}\frac{\sinh^2\;[2\kappa^{\frac{1}{2}}(1-E/\Phi_0)^{\frac{1}{2}}]}{(1-E/\Phi_0)\,E/\Phi_0}\right\}^{-1} \quad \text{for } E<\Phi_0. \tag{3.3.12}$$

* In this case, however, one is concerned with an attractive potential which should be compared with a potential well rather than with a barrier [see problem 3.3.4].

One notes that, for $E \to \Phi_0$, (3.3.12) gives the same value as (3.3.11), namely $T = 1/(1+\kappa)$. The transmission coefficient thus varies continuously when E passes the classical critical value Φ_0. For $E \ll \Phi_0$ and $\kappa \gg 1$ one finds the simple expression

$$T \approx 16 \frac{E}{\Phi_0} \exp(-4\sqrt{\kappa}) \approx 16 \frac{k^2}{\beta^2} \exp(-2\beta b), \qquad (3.3.13)$$

which is seen to decrease exponentially with increasing width of the barrier.

The appearance of a finite transmission for $E < \Phi_0$ is called the tunnel effect. The effect explains the fact that an electric current is able to pass through the boundary surface between two pieces of metal which are brought into contact. It is also important for the understanding of the decay of radioactive nuclei, and of the fact that nuclear reactions can be produced by the bombardment with positively charged particles which, according to classical mechanics, would have an energy too low to overcome the Coulomb repulsion from the nucleus.

It appears from the expressions (3.3.11) and (3.3.12) for the transmission coefficient for a one-dimensional barrier that, when E is measured in units of Φ_0, the transmission properties of the barrier are completely characterized by the dimensionless quantity κ which is given by (3.3.10). Clearly the penetrability of the barrier decreases with increasing κ. Fig. 3.6 shows the transmission coefficient as a function of E/Φ_0 for a barrier with $\kappa = 4$.

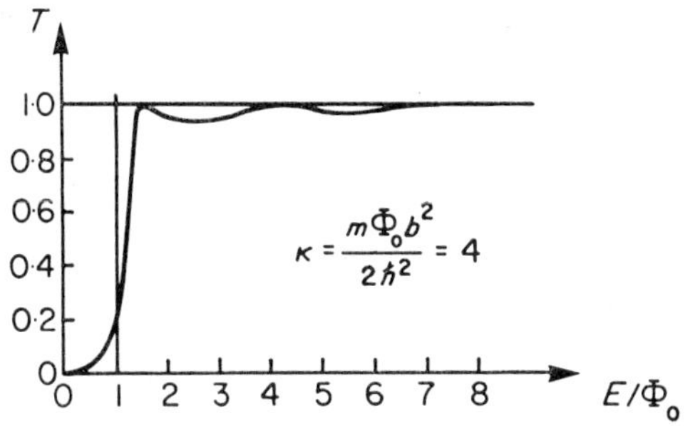

Fig. 3.6

Problems

3.3.1 An electron with kinetic energy 1eV impinges perpendicularly on a metal surface. The potential energy for the electron is assumed to be

-20 eV inside the metal. Find the probability of the reflexion of the electron from the surface.

3.3.2 In Landau and Lifshitz, *Quantum Mechanics*, Pergamon Press, London, 1958 (page 75) the reflexion from a 'soft potential wall' is discussed, given by $\Phi(x) = \Phi_0[1+\exp(-x/\lambda)]^{-1}$ (see Fig. 3.7).

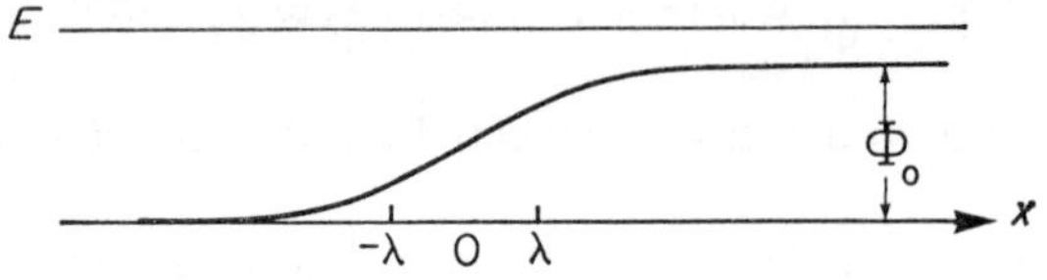

Fig. 3.7

The reflexion coefficient is found to be

$$R = \left\{\frac{\sinh[\pi\lambda(k_1-k_2)]}{\sinh[\pi\lambda(k_1+k_2)]}\right\}^2, \quad \text{where} \quad \begin{aligned} k_1 &= (2mE/\hbar^2)^{\frac{1}{2}} \\ k_2 &= [2m(E-\Phi_0)/\hbar^2]^{\frac{1}{2}}. \end{aligned}$$

Examine for fixed values of E and Φ_0 the significance of the steepness of the potential wall; show that for $\lambda\to0$ the reflexion becomes the same as for a potential wall, and show that $\lambda\to\infty$ leads to the classical result $R = 0$.

3.3.3 A particle passing over the barrier in Fig. 3.5 ($E>\Phi_0$) encounters two subsequent potential walls. Let us set the two equally large transmission coefficients of the potential walls equal to T_s.

If the barrier is wide in comparison with the wave packet which describes the particle, it must be possible to consider the passage over the two potential walls as independent events. Since a transmission through the barrier is the result of transmission at both potential walls, the probability for a barrier transmission must be $T = T_s^2$. This is apparently inconsistent with the expressions (3.3.4) and (3.3.9). Explain the paradox!

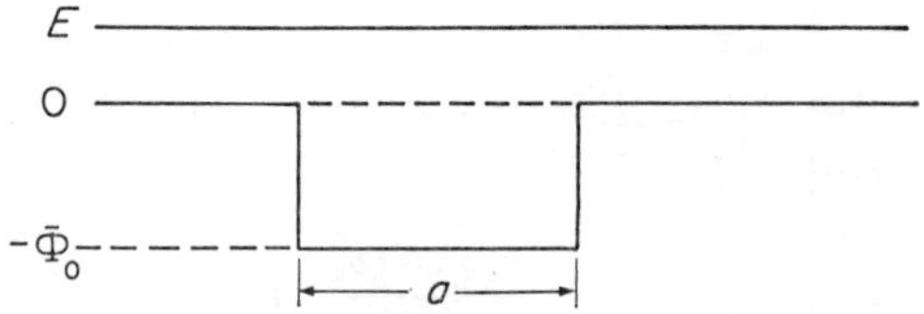

Fig. 3.8

3.3.4 Find the transmission coefficient for a particle of energy E through a one-dimensional potential well of depth Φ_0 and width a (see Fig. 3.8). Especially examine the transmission for $\Phi\ll E$ and $\Phi_0\gg E$. Show that when the well is formed in such a way that it is just unable to contain two energy levels, then one has $T = 1$ for $E\to0$. Plot T for this well as function of E/Φ_0.

Indicate a general condition for a well to give 100% transmission for extremely low energies.

3.4 CONSTANTS OF THE MOTION

In classical mechanics the concept of constants of the motion plays an important rôle. The constants of motion are physical quantities which, in a given mechanical problem, assume values which do not depend on time but only on the initial conditions. In quantum mechanics a physical quantity q is represented by an operator, $\hat{q}$, which can be ascribed a definite value only under very special initial conditions (that ψ be an eigenstate of q). Hence the natural definition of a constant of the motion in quantum mechanics is a physical quantity with a constant expectation value, i.e.

$$\frac{\mathrm{d}}{\mathrm{d}t}\langle\hat{q}\rangle = 0. \tag{3.4.1}$$

It follows from (3.1.4) that (3.4.1) is equivalent to

$$\langle[\hat{q}, \hat{H}]\rangle = 0$$

which is fulfilled for any state only if $\hat{q}$ commutes with the Hamiltonian operator, i.e.

$$[\hat{q}, \hat{H}] = 0. \tag{3.4.2}$$

We mention here as an example that for conservative systems, which are the only ones considered so far, the energy is a constant of the motion. We have shown this already in 1.6, but it also follows directly from (3.4.2) since obviously $\hat{H}$ commutes with itself.

As another example we can mention that for a free particle the three components of momentum are constants of the motion. This follows from the theorem of Ehrenfest, but can also be seen directly from (3.4.2). In 2.1 we saw that the plane waves (2.1.12) were eigenfunctions not only of $\hat{H}$ but also of $\hat{\mathbf{p}}$. Thus in the plane wave states both the energy and the momentum vectors have definite and constant values in time. This is no accident but a consequence of a general theorem which states that two commuting operators have common eigenfunctions.

We will now prove this theorem and thus assume that two operators $\hat{S}$ and $\hat{q}$ satisfy the equation

$$[\hat{S}, \hat{q}] = \hat{S}\hat{q} - \hat{q}\hat{S} = 0. \tag{3.4.3}$$

We imagine that first we determine all eigenfunctions, $u_i(\mathbf{r})$, of the operator $\hat{S}$, i.e. a complete orthogonal system of functions which satisfy the eigenvalue equations

$$\hat{S}u_i(\mathbf{r}) = S_i u_i(\mathbf{r}), \tag{3.4.4}$$

where $S_1, S_2, \cdots, S_i \cdots$ are the eigenvalues.

We now assume that there exists, for each eigenvalue S_i only one eigenfunction, i.e. we assume that the spectrum is non-degenerate. We then let the operator $\hat{q}$ act on both sides of equation (3.4.4) and find, according to (3.4.3)

$$\hat{q}\hat{S}u_i = \hat{S}\hat{q}u_i = S_i\hat{q}u_i. \tag{3.4.5}$$

The latter equation shows that the function $\hat{q}u_i$ is an eigenfunction of the operator $\hat{S}$ belonging to the eigenvalue S_i, but since we assumed that this eigenfunction was uniquely determined, the function $\hat{q}u_i$ must be equal to a constant, q_i times u_i, i.e.

$$\hat{q}u_i(\mathbf{r}) = q_i u_i(\mathbf{r}). \tag{3.4.6}$$

It follows from this equation that u_i is also an eigenfunction of $\hat{q}$ which is just what we wanted to prove.

We will not carry through the proof in detail in the more general case where the spectrum (3.4.4) is degenerate. In this case we must assume that for each eigenvalue S_i there exists a number of linearly independent solutions of (3.4.4). Any arbitrary solution chosen from these may be written as a linear combination of a number of mutually orthonormal eigenfunctions, $u_{il}(\mathbf{r})$, of (3.4.4) with $l = 1, 2, \ldots, n_i$, where n_i indicates the degree of degeneracy. It should be noted that one has a freedom in the choice of this orthogonal system, since one is able to form equivalent orthogonal systems by performing an arbitrary linear orthogonal transformation of the given orthogonal system. One can show now that it is possible to choose this orthogonal system in such a way that the basis functions $u'_{il}(\mathbf{r})$ in the new system are eigenfunctions of $\hat{q}$, i.e.

$$\hat{q}u_{il}(\mathbf{r}) = q_i u'_{il}(\mathbf{r}). \tag{3.4.7}$$

The functions $u'_{il}(\mathbf{r})$ are then the simultaneous eigenfunctions for $\hat{S}$ and $\hat{q}$. The detailed proof of (3.4.7) is left to the reader.

It is now easy to extend this theorem and show that it is possible to find complete sets of common eigenfunctions for three or more operators, provided that all the operators commute mutually.

This theorem plays an important part in quantum mechanics, and in the following we will see many examples of its application. As an example from the previous sections we return to the free particle. In this case the eigenfunctions of the Hamiltonian are highly degenerate. However, since the momentum operators $\hat{p}_x$, $\hat{p}_y$, and $\hat{p}_z$ commute with $\hat{H}$, and since, furthermore, all three momentum operators commute with each other, it is possible, as we saw in (2.1.15), to choose eigenfunctions of $\hat{H}$ which are also eigenfunctions of $\hat{\mathbf{p}}$. We note that the common eigenfunctions are uniquely determined.

Whereas constants of the motion as energy and momentum, are also constants of motion for the corresponding classical cases, we see from (3.4.2) that it may be possible to find operators which are quantum mechanical constants of the motion but which do not have any direct classical analogue. As an example we consider the motion of a particle in a potential which is reflexion-symmetric, which means

$$\Phi(\mathbf{r}) = \Phi(-\mathbf{r}). \tag{3.4.8}$$

In this case one sees that the potential energy commutes with the so-called parity operator $\hat{P}$, which is defined by the equation

$$\hat{P}f(\mathbf{r}) = f(-\mathbf{r}), \tag{3.4.9}$$

where f is an arbitrary wave function. As $\hat{P}$ commutes also with the Laplacian, i.e.

$$\hat{P}\Delta f(\mathbf{r}) = \Delta \hat{P} f(\mathbf{r}), \tag{3.4.10}$$

one finds that $\hat{P}$ is a constant of the motion:

$$[\hat{P}, \hat{H}] = 0. \tag{3.4.11}$$

From this we can conclude that the stationary states (the eigenfunctions of $\hat{H}$) can be chosen such that they are also eigenfunctions of $\hat{P}$, i.e.

$$\hat{P}u = \alpha u, \tag{3.4.12}$$

where α is a constant. It is now easy to determine the eigenvalues α. From the fact that the operator $\hat{P}^2$ is equal to the identity operator, we find

$$\begin{aligned} \hat{P}^2 u &= u \\ &= \hat{P}\alpha u \\ &= \alpha^2 u, \end{aligned} \tag{3.4.13}$$

which shows that

$$\alpha = \pm 1. \tag{3.4.14}$$

The eigenfunctions of $\hat{P}$ are thus even or odd functions. One says that the parity is positive and even, or negative and odd, respectively.

Hence, if the energy spectrum is non-degenerate, we can conclude that the stationary wave functions in reflexion-symmetric potentials are either even or odd functions. This is in accordance with the results found in 2.2–2.4.

In cases where the energy spectrum is degenerate one may choose to describe the stationary states by even and odd functions. For the free particle one can thus use real functions (2.1.14), which have a certain parity. The eigenstates (2.1.12) of the momentum operators can, however, not be chosen as eigenfunctions of $\hat{P}$, which is directly related to the fact that the parity and the momentum operators do not commute.

Finally, we observe that the existence of the uncertainty relations between position and momentum can be related to the fact that the operators x and $\hat{p}_x$ do not commute. One sees easily that the operators satisfy the commutation relation

$$[x, \hat{p}_x] = i\hbar, \tag{3.4.15}$$

and this shows that it is not possible to find states where both x and $\hat{p}_x$ have definite values.

3.5 THE DOUBLE WELL

While the motion of a particle in the potential well in 2.3 can be taken as a rough one-dimensional model of an atom or a nucleus, and the discussion in 3.3 can be taken as a corresponding model for the scattering of electrons on atoms, we investigate in this section a one-dimensional model of a molecule.

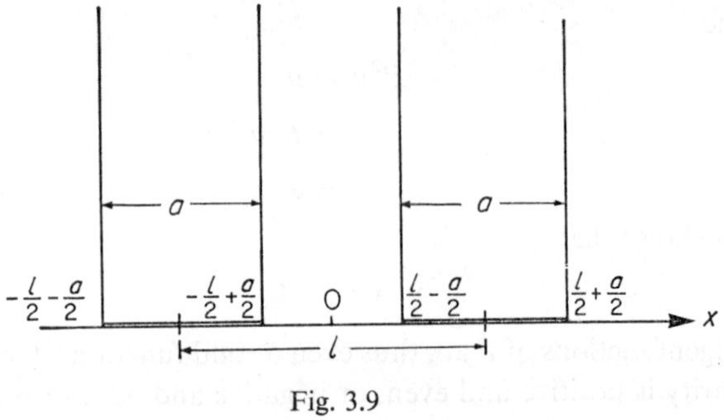

Fig. 3.9

In the crudest first approximation we imagine that the potential energy for an electron, bound in an atom, is given by the potential box in 2.3. The motion of an electron in a molecule consisting of two identical atomic cores, such as $H_2{}^+$, is then represented by the potential shown in Fig. 3.9, where

$$\Phi(x) = \begin{cases} 0, \text{ for } \frac{l}{2}-\frac{a}{2} < |x| < \frac{l}{2}+\frac{a}{2} \\ \infty, \text{ otherwise.} \end{cases} \tag{3.5.1}$$

In this potential we can at once write down a set of solutions to the stationary Schrödinger equation by using the results in 2.2:

$$u_n^{(r)}(x) = A_n \begin{cases} \cos \\ \sin \end{cases} \left\{k_n\left(x-\frac{l}{2}\right)\right\}, \text{ for } \left|x-\frac{l}{2}\right| < \frac{a}{2} \\ \quad 0 \quad \text{,otherwise,} \tag{3.5.2}$$

with

$$k_n = n\frac{\pi}{a} \quad (n = 1, 2, 3, \cdots) \tag{3.5.3}$$

and

$$E_n = \frac{\hbar^2 k_n{}^2}{2m}. \tag{3.5.4}$$

The solutions (3.5.2)–(3.5.4) represent states for the electron in the atom on the right hand side of Fig. 3.9, since the probability amplitude is only different from zero in this potential box. There also exists, however, a series of states corresponding to the presence of the electron in the atom on the left-hand side:

$$u_n^{(l)}(x) = A_n \begin{cases} \cos \\ \sin \end{cases} \left\{k_n\left(x+\frac{l}{2}\right)\right\}, \text{ for } \left|x+\frac{l}{2}\right| < \frac{a}{2} \\ \quad 0 \quad \text{,otherwise.} \tag{3.5.5}$$

These states can be derived from (3.5.2) by substituting $x-l/2$ by $x+l/2$. One notices that the states are orthogonal to the states (3.5.2). The energy values for the new states are exactly the same as before, namely (3.5.4), and the resulting energy spectrum is therefore the same as the spectrum (Fig. 2.3) for one atom, with the only difference that each state is doubly degenerate. This result is not surprising. We have assumed the potential to be infinite outside the atoms, and we have thereby forced the electrons to be either at one atom or at the other. In fact we must expect that this is only true when the atoms are separated by a large distance. In a more realistic model with finite potentials, we must, according to 3.2, take into consideration that the electron is able to penetrate the barrier between the atoms; a state like (3.5.2), where the electron is inside one of the atoms, is then no longer a stationary state. The value of the potential outside the atoms ($|x| > l/2+a/2$) plays a minor rôle for the bound electrons, and in order to simplify the calculations as much as possible we will only improve the model by making the potential between the atoms finite. Hence we assume

$$\Phi(x) = \begin{cases} \Phi_0, \text{for } |x| < \dfrac{l}{2}-\dfrac{a}{2} \\ 0, \text{ for } \dfrac{l}{2}-\dfrac{a}{2} < |x| < \dfrac{l}{2}+\dfrac{a}{2} \\ \infty, \text{for } |x| > \dfrac{l}{2}+\dfrac{a}{2}. \end{cases} \tag{3.5.6}$$

We must now seek solutions to the stationary Schrödinger equation in each of the three regions. With the boundary conditions $u = 0$ at $|x| = l/2+a/2$, the solutions are

$$\begin{aligned} u_{\text{I}} &= A \sin k\left(x+\frac{a}{2}+\frac{l}{2}\right), && -\frac{l+a}{2} < x < \frac{a-l}{2} \\ u_{\text{II}} &= B \exp(\beta x)+C \exp(-\beta x), && |x| < \frac{l-a}{2} \\ u_{\text{III}} &= D \sin k\left(x-\frac{a}{2}-\frac{l}{2}\right), && \frac{l-a}{2} < x < \frac{l+a}{2} \end{aligned} \tag{3.5.7}$$

where

$$k = \left(\frac{2mE}{\hbar^2}\right)^{\frac{1}{2}}, \qquad \text{and} \quad \beta = \left(\frac{2m(\Phi_0-E)}{\hbar^2}\right)^{\frac{1}{2}}.$$

We now assume that $E<\Phi_0$, and from the continuity conditions at $x = \pm(l-a)/2$ we find four equations to determine the constants

A, B, C and D. By calculation one finds that these equations can only be satisfied if either

(i) $$B = C \quad \text{and} \quad \frac{k}{\beta}\cot ka = -\tanh \beta\left(\frac{l}{2}-\frac{a}{2}\right), \tag{3.5.8}$$

or

(ii) $$B = -C \quad \text{and} \quad \frac{k}{\beta}\cot ka = -\coth \beta\left(\frac{l}{2}-\frac{a}{2}\right). \tag{3.5.9}$$

In the case (i) one finds furthermore

$$A = D = C\frac{2\cosh\,(\beta\,(l-a)/2)}{\sin ka}, \tag{3.5.10}$$

while in the case (ii) one finds

$$A = -D = C\frac{2\sinh\,(\beta\,(l-a)/2)}{\sin ka}. \tag{3.5.11}$$

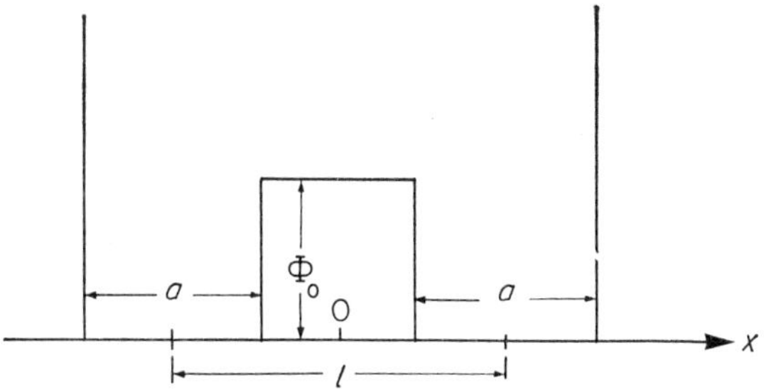

Fig. 3.10

The equations (3.5.8) and (3.5.9) determine the energy spectrum. If the distance l between the atoms is very large, one sees that the two equations lead to identical eigenvalues, i.e. to a degenerate spectrum determined by

$$k\cot ka = -\beta. \tag{3.5.12}$$

This situation is analogous to the case discussed above ($\Phi_0 \to \infty$), where the electron is not able to penetrate the barrier and the spectrum (3.5.12) is that of the free atom.

For finite values of l and Φ_0 the degenerate states are split up in two states, as shown in Fig. 3.11. We notice that the equation (3.5.8) leads to an energy E_+, which is lower than the energy (3.5.12) of the 'free

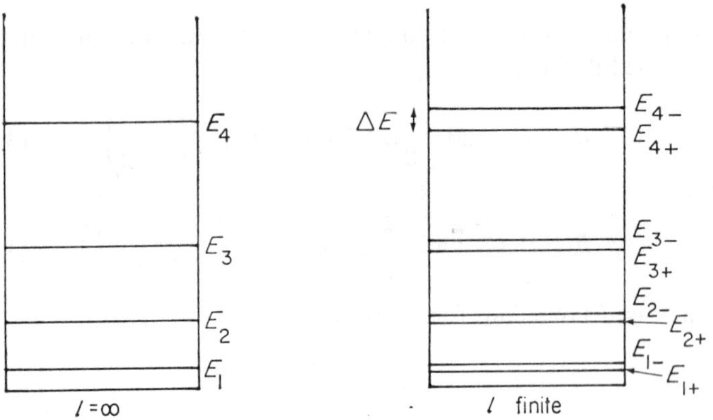

Fig. 3.11

atom', while (3.5.9) leads to a higher energy E_-. To first order the difference between these energies and thc energy of the corresponding state E for the free atom is given by

$$|E_+ - E| \approx 4\,\frac{(\Phi_0 - E)E}{\Phi_0(1+\beta a)}\exp\,[-\beta(l-a)]. \tag{3.5.13}$$

When the molecule is in its ground state one must supply energy in order to separate it into free atoms. This binding of the atoms is a purely quantum mechanical effect due to the fact that through the tunnel effect the electrons in molecules have the possibility of moving around in a larger region of space.

We note that the stationary wave functions which we have just found are even and odd functions in the cases (i) and (ii) respectively. Since the spectrum is non-degenerate we are, according to 3.4, able to predict this solely from the fact that the potential (3.5.6) is symmetrical. When one finds, in the limit $\Phi_0 \to \infty$ (or $l \to \infty$), eigenstates (3.5.2) and (3.5.5), which are neither even nor odd functions, this is possible only because of the degeneracy of the spectrum in this limit. We could also in this case choose the eigenfunctions in such a way that they were eigenfunctions of the parity operator. Thus we could use, instead of (3.5.2) and (3.5.5), the orthogonal wave functions

$$u_+(x) = u^l(x) + u^r(x) \tag{3.5.14}$$

and

$$u_-(x) = u^l(x) - u^r(x), \tag{3.5.15}$$

for which $\hat{P}u_{\pm} = \pm u_{\pm}$. These wave functions represent, for large Φ_0, a very good approximation to the exact wave functions (3.5.7)–(3.5.11).

Finally, we mention that one may imagine a situation where one has proved by an experiment that the electron is, for example, in the left-hand atom. In that case the wave function is, at the time $t = 0$ when the measurement was made, of the type (3.5.5). This wave function is approximately given by the equation

$$\psi(x, t = 0) = u^l(x) \approx u_1 + u_2, \tag{3.5.16}$$

where u_1 is the symmetrical solution (i) given by (3.5.8) and (3.5.10), while u_2 is the corresponding antisymmetrical solution (ii) given by (3.5.9) and (3.5.11).

Of course, the state described by (3.5.16) is not stationary. At a later time t one finds, according to 2.5, that the wave function is given by

$$\psi(x, t) = \exp\left(-\frac{i}{\hbar} E_+ t\right)\left[u_1 + u_2 \exp\left(\frac{i}{\hbar}(E_+ - E_-)t\right)\right]. \tag{3.5.17}$$

If l is large, the energy difference $E_- - E_+$ [compare (3.5.13)] is small, and the probability distribution is therefore unchanged during a long period $t < \hbar/(E_- - E_+)$. The state is quasi-stationary. After the period $t = \pi\hbar/(E_- - E_+)$ the sign of last term in (3.5.17) changes and the wave function then represents a state where the electron is in the right-hand atom. The electron will thus periodically be exchanged between the atoms. The frequency of exchange is given by $\nu = (E_- - E_+)/h$, and one notices that this frequency is a measure of the energy with which the atoms are bound together.

3.6 PERIODIC POTENTIAL. ENERGY BANDS

In this section we will—again on the basis of a one-dimensional model —study effects which are characteristic for electrons in a solid, or more accurately a crystal lattice. The innermost electrons in the atoms of the crystal lattice are, of course, only to a small degree affected by the fact that the atom is not completely isolated. One sees however, from the ability of the metals to conduct electric current, that in some materials there must exist electrons which are able to move through the entire crystal lattice. We shall study the motion of these electrons.

As a starting point we will consider the idealized case, a perfect crystal, where all of the atoms are situated exactly at the lattice points. Because of thermal motion and impurity atoms in the lattice it is not possible to realize the perfect structure, but it can be approached at very low temperatures and with a high degree of chemical purity. For electrons moving through a perfect crystal the potential* is a function which varies periodically in space, and we will, in the following one-dimensional model, examine the motion of a particle in a potential of this kind. We will find the characteristic properties of the energy spectrum and study the motion of a wave packet through the lattice.

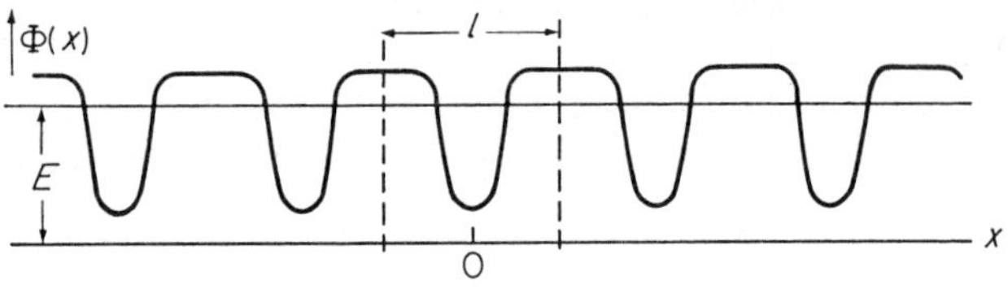

Fig. 3.12

Let the potential be periodic with the period l as indicated in Fig. 3.12. It is assumed to extend over the entire x-axis which corresponds to an infinitely large crystal. The individual potential wells represent the atoms in the one-dimensional lattice.

The stationary Schrödinger equation for a particle in this potential is

$$-\frac{\hbar^2}{2m}\frac{d^2u}{dx^2}+\Phi(x)u = Eu, \tag{3.6.1}$$

where $\Phi(x+l) = \Phi(x)$.

Even without knowledge of the explicit form of the potential $\Phi(x)$ one can—merely from the periodicity—draw certain conclusions as to the possible types of solutions to (3.6.1). If in this equation one replaces x by $x+l$ the potential remains the same, and so does the

* In fact the problem under discussion is a complicated many-body problem, but to a certain approximation one may treat the motion of the individual electron in the crystal as a one-particle problem. The potential is then determined by the average field from the nuclei and from the other electrons in the crystal (compare 5.5).

kinetic energy. In other words, the linear operator $\hat{\Gamma}$, defined by $\hat{\Gamma}f(x) = f(x+l)$ for any function $f(x)$, commutes with the Hamiltonian:

$$[\hat{\Gamma}, \hat{H}] = 0.$$

According to 3.4 the physical quantity corresponding to the displacement operator $\hat{\Gamma}$ is then a constant of the motion. Like the parity it does not have any direct classical analogy, but from the results in 3.4 we can conclude that the stationary states can be chosen so that the wave functions are eigenfunctions of both $\hat{H}$ and $\hat{\Gamma}$. It is thus possible to choose two independent solutions $u_1(x)$ and $u_2(x)$ of (3.6.1), which have the simple property

$$\hat{\Gamma}u_1(x) = u_1(x+l) = \lambda_1 u_1(x)$$

$$\hat{\Gamma}u_2(x) = u_2(x+l) = \lambda_2 u_2(x). \tag{3.6.2}$$

The eigenvalues λ_1 and λ_2 are easily seen to be the reciprocal of each other, because the Wronski determinant,

$$W(x) = u_1(x)u_2'(x) - u_2(x)u_1'(x),$$

is constant (and $\neq 0$) on the entire x-axis, and from (3.6.2) one finds

$$W(x+l) = \lambda_1\lambda_2 W(x),$$

and consequently

$$\lambda_1\lambda_2 = 1. \tag{3.6.3}$$

Moreover, one finds directly by complex conjugation of the equations (3.6.2) that λ_1^* and λ_2^* are again eigenvalues of $\hat{\Gamma}$. These eigenvalues must, however, be the same as the old ones, and we therefore have the following alternative. We must either have

$$\lambda_1^* = \lambda_1$$

$$\lambda_2^* = \lambda_2,$$

so that λ_1 and λ_2 are real, or we must have

$$\lambda_1^* = \lambda_2$$

$$\lambda_2^* = \lambda_1,$$

so that $\lambda_1^*\lambda_1 = \lambda_2^*\lambda_2 = 1$. Only in the latter case will the wave functions remain finite for $x \to +\infty$ as well as for $x \to -\infty$.

For acceptable wave functions the eigenvalues λ_1 and λ_2 are thus of the form exp (iα) and exp ($-$iα), where α is real, i.e.

$$u(x) = \exp(ikx)\phi(x), \tag{3.6.4}$$

where k is real and $\phi(x+l) = \phi(x)$. The two independent solutions of this type will change by a factor exp (ikl) and exp ($-ikl$), respectively, when one moves from x to $x+l$. The solutions (3.6.4) for the periodic potential are called the Bloch waves.

In order to show in actual cases, how one may obtain solutions of the form (3.6.4), the simplifying assumption is made that $\Phi(x)$, as indicated in Fig. 3.13, is symmetric around the point $x = 0$, which corresponds to the minimum of a potential well. We choose the interval $-l/2<x<l/2$ as the basic domain; here the general solution can be written in the form $u(x) = Af(x)+Bg(x)$, where $f(x)$ and $g(x)$ are real functions which, due to the symmetry of the potential, may

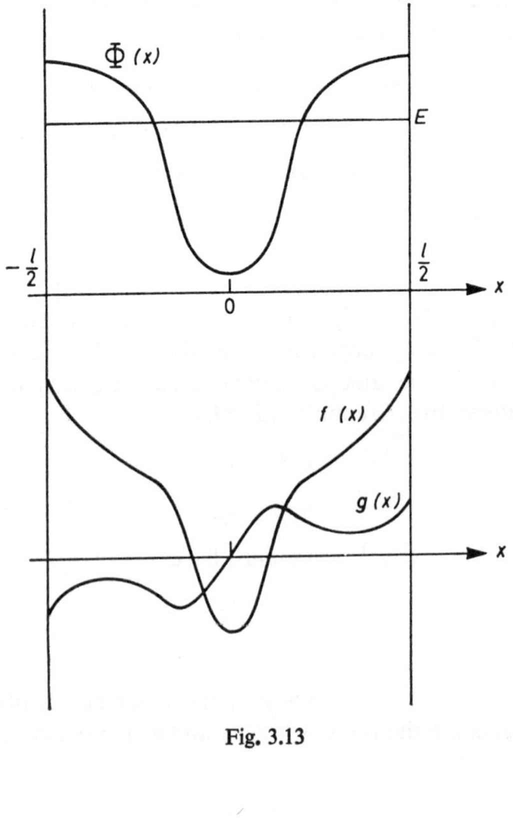

Fig. 3.13

be assumed to be respectively even and odd functions of x (compare 3.4). This is shown in Fig. 3.13.

According to (3.6.4), both $u(x)$ and $u'(x)$ should, in the interval $l/2$ to $3l/2$, result from the corresponding quantities at the point $x-l$ (i.e. in the basic domain) through a multiplication by exp (ikl). The continuity conditions at $x = l/2$ are thus

$$Af\left(\frac{l}{2}\right)+Bg\left(\frac{l}{2}\right) = \exp(ikl)\left[Af\left(-\frac{l}{2}\right)+Bg\left(-\frac{l}{2}\right)\right],$$

$$Af'\left(\frac{l}{2}\right)+Bg'\left(\frac{l}{2}\right) = \exp(ikl)\left[Af'\left(-\frac{l}{2}\right)+Bg'\left(-\frac{l}{2}\right)\right].$$

Using the parity of $f(x)$ and $g(x)$ one gets

$$Af\left(\frac{l}{2}\right)[1-\exp(ikl)]+Bg\left(\frac{l}{2}\right)[1+\exp(ikl)] = 0$$

$$Af'\left(\frac{l}{2}\right)[1+\exp(ikl)]+Bg'\left(\frac{l}{2}\right)[1-\exp(ikl)] = 0,$$

which only has proper solution for A and B if

$$\cot\frac{kl}{2} = \left(-\frac{f}{f'}\frac{g'}{g}\right)^{\frac{1}{2}}.$$

This condition may be written

$$\cos kl = \frac{fg'+gf'}{fg'-gf'}, \tag{3.6.5}$$

where all values of the functions should be taken at the point $x = l/2$.

In an actual case one can find for each energy value E, the even and odd solutions in the basic domain and thereafter calculate the right-hand side of (3.6.5). As we shall see presently the right-hand side of (3.6.5) is an oscillating function of E and the energy spectrum will then consist of the energies in the intervals where this function lies between $+1$ and -1. For energies outside these intervals, (3.6.5) can only be satisfied for imaginary values of k, and according to the previous discussion this leads to solutions which do not remain finite when $x\to\pm\infty$.

Before we proceed with the general investigation we will illustrate the discussion above by determining the energy spectrum in a potential, the so-called Kronig–Penney model, which leads to elementary calculations. Let the lattice be composed of rectangular potential wells of width a and barriers of width $l-a$ and height Φ_0 (see Fig. 3.14).

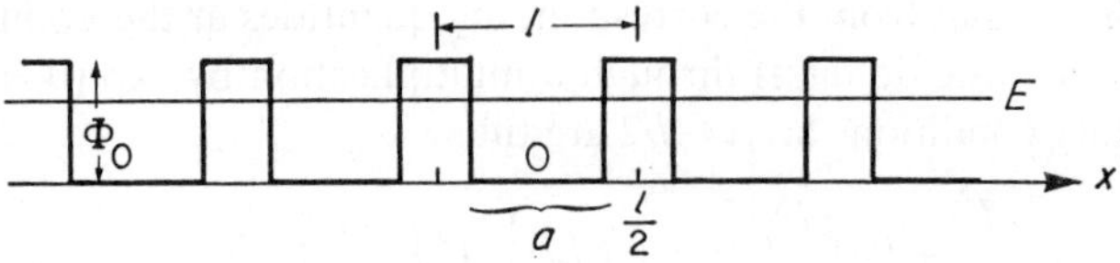

Fig. 3.14

Consider first the case $E<\Phi_0$, and introduce the real constants

$$\alpha = \left(\frac{2mE}{\hbar^2}\right)^{\frac{1}{2}}, \qquad \beta = \left(\frac{2m(\Phi_0-E)}{\hbar^2}\right)^{\frac{1}{2}}. \tag{3.6.6}$$

In the basic region we can then use solutions

$$\begin{aligned} f(x) &= \cos \alpha x \\ g(x) &= \sin \alpha x \end{aligned} \qquad \text{for } |x|<\frac{a}{2}. \tag{3.6.7}$$

Using the continuity conditions at the point $x = a/2$, we find that on the right-hand side of $x = a/2$ these solutions are continued into

$$\begin{aligned} f(x) &= \cos\left(\alpha\frac{a}{2}\right)\cosh\left(\beta\frac{x-a}{2}\right)-\frac{\alpha}{\beta}\sin\left(\alpha\frac{a}{2}\right)\sinh\left(\beta\frac{x-a}{2}\right) \\ g(x) &= \sin\left(\alpha\frac{a}{2}\right)\cosh\left(\beta\frac{x-a}{2}\right)+\frac{\alpha}{\beta}\cos\left(\alpha\frac{a}{2}\right)\sinh\left(\beta\frac{x-a}{2}\right). \end{aligned} \tag{3.6.8}$$

We now know the even and the odd solution at $x = l/2$, and we are able to calculate the right-hand side of (3.6.5). One sees immediately that the denominator in this expression is the Wronski determinant, which can be calculated at the point $x = 0$ and gives the result α. The numerator is the derivative of the product $f \cdot g$, taken in the point $x = l/2$. For the product, one finds from (3.6.8) that

$$\begin{aligned} f(x)g(x) &= \tfrac{1}{2}\frac{\alpha}{\beta}\cos(\alpha a)\sinh[\beta(2x-a)] \\ &\quad +\tfrac{1}{2}\sin(\alpha a)\left[1+\left(1-\frac{\alpha^2}{\beta^2}\right)\sinh^2\frac{\beta(2x-a)}{2}\right], \end{aligned}$$

which by differentiation and insertion of $x = l/2$ yields for the condition (3.6.5)

$$\cos kl = \cos \alpha a \cosh \beta(l-a) + \frac{\beta^2 - \alpha^2}{2\alpha\beta} \sin \alpha a \sinh \beta(l-a). \tag{3.6.9}$$

The corresponding condition for $E > \Phi_0$ can be found from (3.6.9) by substituting β by $i\beta'$, where $\beta' = [2m(E-\Phi_0)/\hbar^2]^{\frac{1}{2}}$. Thus one gets

$$\cos kl = \cos \alpha a \cos \beta'(l-a) - \frac{\beta'^2 + \alpha^2}{2\alpha\beta'} \sin \alpha a \sin \beta'(l-a). \tag{3.6.10}$$

The right-hand side of (3.6.9) and (3.6.10) are, for given values of $(l-a)/a$ and $\kappa = m\Phi_0 a^2/2\hbar^2$, a function of E/Φ_0. Fig. 3.15 shows this function and the corresponding energy spectrum for $\kappa = 36$ and $(l-a)/a = 1/24$.

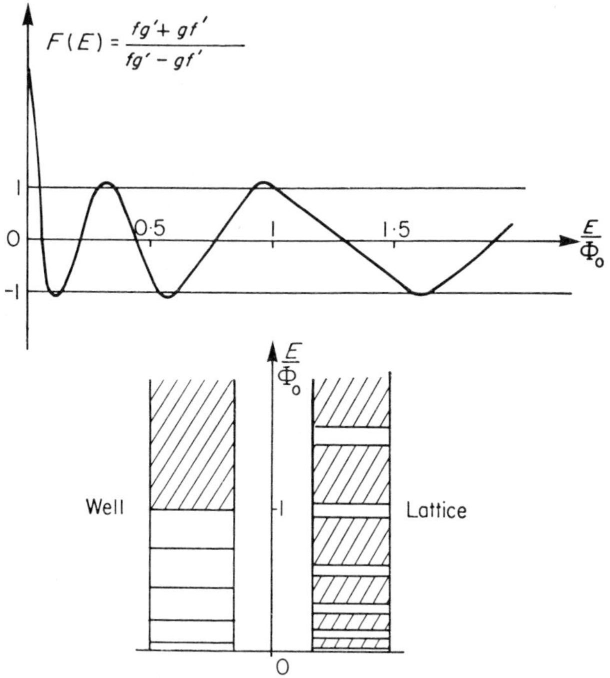

Fig. 3.15

As a comparison, the energy spectrum for one of the potential wells, of which the lattice is composed, is also illustrated. It should be noted that the energy spectrum for the lattice consists of bands where all energies are allowed alternating with forbidden bands. For $E<\Phi_0$ each of the allowed bands correspond to a discrete level in the well and one can show [problem 3.6.1] that with increasing distance between the wells in the lattice the bands become narrower and narrower until they finally coincide with the levels (compare also 3.5). In the region $E>\Phi_0$ it is remarkable that forbidden bands appear. We shall discuss their significance later.

Qualitatively, the band structure found is not limited to the special potential discussed above. This can be seen by considering the general condition (3.6.5) which may be rewritten in the form

$$\cos kl = \frac{(f'/f)+(g'/g)}{(f'/f)-(g'/g)} \tag{3.6.11}$$

where it contains only the logarithmic derivatives of the solutions f and g on the boundary of the basic region. From (3.6.11) one sees that f'/f and g'/g must have opposite signs in the allowed bands. This is, however, exactly the case for energies near the discrete levels in the individual potential wells. For these energies either f or g will be the correct wave function, which tends asymptotically towards zero so that the corresponding logarithmic derivative is negative, while the other is positive because the solution tends rapidly towards infinity. On the other hand, far away from the discrete levels, both f and g tend rapidly towards infinity so that both f'/f and g'/g are positive and (3.6.11) cannot be satisfied at such energies.

We will now briefly discuss the stationary wave functions belonging to the energies in the allowed bands. For a given energy there will be two independent wave functions of the form (3.6.4), which can be written

$$\exp(ikx)\,\phi_k(x) \quad \text{and} \quad \exp(-ikx)\,\phi_{-k}(x), \tag{3.6.12}$$

where we have emphasized the fact that the function $\phi(x)$ depends on k. It is noted that $\phi_{-k} = \phi_k{}^*$. The value of k is connected with the energy through (3.6.5) or (3.6.11) ,where k only enters through $\cos kl$. One sees, however, since $\cos kl$ as a function of k is periodic with the period $2\pi/l$, that (3.6.5) is satisfied for the same energy value by infinitely many other k-values, and according to the derivation of

(3.6.5) it is possible to use all these values in (3.6.12). In this way, however, one does not achieve new wave functions because the differential equation (3.6.1) can have no more than two independent solutions. The explanation is that when k is changed by an integral multiple of $2\pi/l$, $\phi_k(x)$ will change in such a way that the resulting wave function remains unchanged. All states in an energy band are apparently described by letting k take on values in an interval of width $2\pi/l$ and it is convenient to use the interval $-\pi/l<k<\pi/l$. When the same k-interval is used for all energy bands one must just keep in mind that the functions ϕ in (3.6.12) will depend—apart from k—also on a label specifying the energy band*.

We found above that there are allowed energy bands, within which all energies are permissible because we considered an infinite lattice.

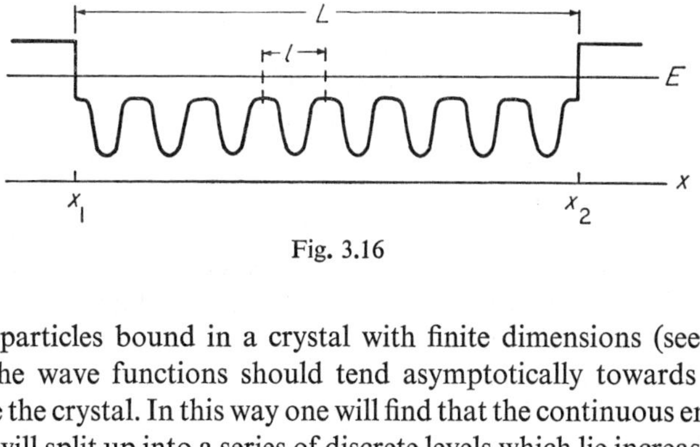

Fig. 3.16

For particles bound in a crystal with finite dimensions (see Fig. 3.16) the wave functions should tend asymptotically towards zero outside the crystal. In this way one will find that the continuous energy bands will split up into a series of discrete levels which lie increasingly close together as the crystal becomes larger. This structure in the energy bands is without importance in most problems and must only be taken into consideration if one is interested in the number of states in an energy band. We may evaluate the number of quantum states within an energy band in a finite lattice, for instance by letting the crystal boundaries in Fig. 3.16 correspond to infinitely high potential walls. We assume, as indicated, that there is an integral number of

* One often uses another representation in which different energy bands correspond to different k-intervals, thus achieving a one-to-one correspondence between the k-value and the state. The interval $|k|<\pi/l$ describes the lowest energy band; the next band is described by $\pi/l<|k|<2\pi/l$, and so forth [see problem 3.6.6].

lattice periods within the boundaries. Inside the crystal the general solution for a given energy has the form

$$v(x) = A \exp(ikx)\, \phi_k(x) + B \exp(-ikx)\, \phi_k^*(x),$$

where A and B are constants.

We demand that $v(x_1) = 0$, which determines the ratio of A and B. Since $\phi_k(x_1) = \phi_k(x_2)$ [and $\phi_k^*(x_1) = \phi_k^*(x_2)$] the wave function at the boundaries is of the form

$$A' \exp(ikx) + B' \exp(-ikx),$$

where $A' = A\phi_k(x_1)$ and $B' = B\phi_k^*(x_1)$. The boundary conditions are thus completely the same as for a particle of wave number k in a potential box of length $L = x_2 - x_1$, and we therefore find that $kL = 2\pi n$, where n is an integer. Thus the allowed k-values lie equidistant along the k-axis with separation $2\pi/L$. Since an energy band corresponds to the k-values from $-\pi/2$ to $+\pi/2$, one sees that the number of states in the band is L/l, which is the same as the number of 'atoms' in the crystal (compare also the double well 3.5).

We now return to the infinite potential lattice, and on the basis of the discussion of the stationary states we will investigate the motion of a particle through the lattice. Here one notices at once a certain similarity between the wave functions (3.6.12) and the harmonic waves describing a free particle. The only difference is that in (3.6.12) the harmonic waves are modulated by a periodic function with the same period as the potential. This modulation will have little influence on the construction of wave packets when these are long in comparison with l: this is the reason why the particle in certain respects behaves as if it were free. A wave packet built up of wave functions (3.6.12) with k-values in a small interval will thus travel through the lattice with a constant velocity, namely the group velocity

$$v_g = \partial\omega/\partial k = \hbar^{-1}\partial E/\partial k.$$

This is seen immediately by the same reasoning as in 1.2, when one assumes that the modulation function can be regarded as independent of k in the small k-interval. A rigorous proof will not be attempted here.

The free passage of a particle through the lattice may seem paradoxical because the particle passes through or over a series of potential barriers which should cause reflexion. The phenomenon is due to the periodicity of the potential, and it has a close analogy in the passage of electromagnetic waves through a crystal. When this is perfect, the waves scattered from the atoms of the lattice interfere in such a way that the waves pass through without any net scattering. Only deviations from the perfect crystal structure will give an actual scattering and thereby a damping of the wave.

By the unhindered passage of an electromagnetic wave through a perfect crystal we disregarded the effects of diffraction appearing at certain wavelengths where it is possible to satisfy the Bragg conditions for the crystal. The corresponding phenomenon appears also for the particle in the one-dimensional lattice, and is due to the forbidden bands. Solutions to the stationary Schrödinger equation will, for energies in these bands, be unacceptable for an infinite crystal because they cannot remain finite for $x \rightarrow \pm \infty$. If, however, the crystal extends only from $x = 0$ to $x = +\infty$, it is possible to accept a solution which is divergent for $x \rightarrow -\infty$ but tends towards zero for $x \rightarrow +\infty$. If this solution is combined with solutions for a free particle for $x < 0$, then one has a stationary description of a particle which moves towards the crystal and is totally reflected (since the current is zero at $x \rightarrow +\infty$). Thus a total reflexion occurs when the energy of the incoming particle corresponds to one of the forbidden bands in the crystal. The discrete wavelengths characterized by Bragg's conditions should actually be substituted by wavelength intervals. This is also brought out by a more accurate analysis and is in agreement with experiment.

We can check the above interpretation of the forbidden bands by considering the special lattice potential of Fig. 3.14. From (3.6.10) one finds, when $E \gg \Phi_0$ and consequently $\beta' \approx \alpha$, the approximate conditions for the allowed bands:

$$\cos kl \approx \cos \alpha a \cos \alpha(l-a) - \sin \alpha a \sin \alpha(l-a) = \cos \alpha l \tag{3.6.13}$$

In this approximation the forbidden bands do not have any width, but they must lie at the boundaries of the allowed bands which, according to (3.6.13), correspond to $\alpha \approx n\pi/l$ (n integer). If the potential jump at the edge of the crystal is assumed to be of the same order of magnitude as Φ_0, one may in the considered energy region ($E \gg \Phi_0$) set the wavelength outside the crystal to be $\lambda \approx 2\pi/\alpha$. Total reflexion then occurs for wavelengths satisfying $n\lambda \approx 2l$, which is exactly the Bragg condition.

The free mobility in the crystal of a particle in the allowed bands plays an important rôle in the theory of conductivity of metals. As in the elementary study of electricity, one gets a classical description of the conduction phenomenon by means of a model where metal electrons are assumed to move freely in all directions in the time between their collisions with the atoms of the lattice. An outer electric field accelerates the electrons between the collisions, and the contribution of each electron to the net current will then be determined by the extra component of velocity in the field direction which can be obtained in average during the time of free flight, τ. It turns out, however, that in order to get a quantitative correspondence with the observed conductivities, one is forced to assume τ to be several orders of magnitude larger than expected on the basis of the velocity of the electrons and their mean free path which, in the classical theory, is determined by the size and density of atoms. This difficulty disappears in the quantum mechanical description where—as in the one-dimensional model—the electrons are not scattered at all when the crystal is perfect. The scattering only occurs by defects in the structure of the crystal and by oscillations of the atoms around their equilibrium positions. The amplitude of these oscillations diminishes with decreasing temperature, and consequently the mean free path and the conductivity both increase. At quite low temperatures the scattering by defects and impurities becomes dominant so that the mean free path becomes constant.

Here we have only mentioned the qualitative significance of the mean free path. In a detailed theory for electrical conductivity one must, in addition to a calculation of the mean free path, also take into account the velocity of an electron, its acceleration in an external field, and the effect of the interaction between the many electrons. It then turns out that one is only with strong reservations able to maintain the classical picture of the electrons in the crystal as free particles. We will conclude this section by mentioning some of these reservations.

Previously, we found that a particle in the one-dimensional lattice could be regarded as free in the sense that a wave packet travels with constant velocity. However, this velocity is given by

$$v_g = \hbar^{-1} \partial E / \partial k,$$

and hence the dependence of v on E will be determined by the connexion between k and E which is usually much more complicated than for a free particle. In order to get an idea of the connexion between k and E in a band, we assume that the function $F(E)$, representing the right-hand side of (3.6.5), varies rapidly with E around the value E_0, where $F(E_0) = 0$ (this corresponds to a narrow energy band). We can then use a series expansion of $F(E)$ and write (3.6.5) in the form

$$\cos kl = F'(E_0)\,(E-E_0)+\cdots\cdots,$$

from which

$$E \approx E_0+\frac{1}{F'(E_0)}\cos kl.$$

This gives the typical relation between E and k which is shown on Fig. 3.17 in the cases $F'(E_0)<0$ (full-drawn) and $F'(E_0)>0$ (dotted).

For neighbouring energy bands, $F'(E_0)$ will have opposite signs as is seen from Fig. 3.15.

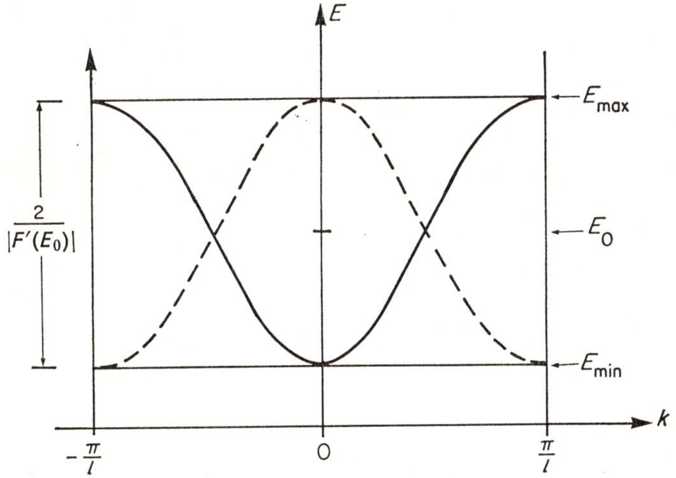

Fig. 3.17

From the above figure one sees that the group velocity, $\hbar^{-1}\partial E/\partial k$, is zero at the bottom and the top of each energy band, and that it does not vary monotonically with the energy through the band. For comparison we note that for a really free particle there is a monotonic relation of the type $v_g = [2(E-\Phi_0)/m]^{\frac{1}{2}}$. Similar complications are found for the acceleration of the electron in an external field. One is able to show that a wave packet in the one-dimensional potential lattice, when acted on by a constant force, K, will have an acceleration $K\hbar^{-2}\partial^2E/\partial k^2$. This leads to the introduction of the quantity

$$m^* = (\hbar^{-2}\partial^2E/\partial k^2)^{-1}$$

as an effective mass for the particle. From the figure above, the effective mass is seen to be positive in the lower part of an energy band but negative in the upper part. Therefore particles in this part of a band will be accelerated in the opposite direction to the field. This is very important in the study of the transport of charge by electrons in a metal, because the electrons in the upper part of an energy band may counteract the transport of charge due to electrons in the lower states. Thus apparently the total transport of charge will depend on the distribution of the electrons in the band. This is determined by the fact that the electrons will assume the lowest* possible energy in accordance with the Pauli principle. As is known from atomic physics, the Pauli principle demands that there are at most two electrons with the same wave function; the energy bands will therefore be filled from the bottom, up to a certain height. If the last band is not completely filled by this process the crystal is a conductor. If, however, the band is just filled up, it is possible to show that the upper electrons in the band will completely compensate the transport of charges by the lower ones [problem 3.6.3] and the material then becomes an insulator. Semiconductors are materials with an empty band slightly above a completely filled band, so that a thermal excitation or a few impurities may raise the electrons to the empty band and make the material a conductor. It is seen that in this case the conductivity increases with temperature, contrary to what is found for normal conductors.

Problems

3.6.1 Examine the condition (3.6.9) in the limit $l \to \infty$, a fixed, and show that the energy bands shrink into levels which agree with the levels found in 2.3 for a well of the width a and depth Φ_0.

3.6.2 Show that close to the bottom (E_{min}) of an energy band, one finds an approximate relation between E and the group velocity v, of the simple form

$$E - E_{min} = \tfrac{1}{2}m^*v^2,$$

where m^* is the effective mass at the bottom of the band.

---

* We will disregard the thermal excitation energy, which is small at normal temperatures in comparison with the width of the band.

3.6.3 In a one-dimensional perfect crystal each state in an energy band is assumed to be occupied by one electron. An electric field of strength $\mathscr{E}$ in the direction of the x-axis acts for a time t, during which the electrons are assumed to be accelerated independently of each other. Find the contribution of the single electron to the resulting current after the time t in terms of the effective mass. Show next, on the basis of the expression for the effective mass and the previous discussion of the distribution of the states in the band, that the total current is zero.

3.6.4 Examine the conditions (3.6.9) and (3.6.10) in the limit $a \to l$, l fixed, and show that the forbidden bands shrink. Introduce the association of k-values with energy bands mentioned in the footnote on page 115, and show that in this limit one can identify k with the wave number α when one is not too close to a forbidden band. Draw from this a qualitative sketch of E as a function of k.

3.7 QUASI-STATIONARY STATES. DECAY

In atomic and nuclear physics one often meets systems which may remain in an apparently unchanged state for a long time and then suddenly change. The change, or decay, may, for instance, occur by emission of electromagnetic radiation or—more drastically—by decomposition of the system. As examples, one may think of excited atoms, radioactive nuclei or mesons. The process may approximately be regarded as a one-particle problem, especially for α-decay, where the α-particle, after a long stay in the nucleus, suddenly breaks through the Coulomb barrier.

As is well known, decay processes follow a statistical law, so that the probability of decay during a short time dt is equal to dt/τ, where τ is a constant called the lifetime (actually the mean lifetime) characteristic for the system. From this empirical law one deduces further that if the system at $t = 0$ is found to be in the initial state, then the probability of finding it unchanged at the time t is given by $\exp(-t/\tau)$.

The initial state for a system which is able to decay often closely resembles a stationary state. It is thus possible for the system to remain unchanged during a long period (for the radioactive nuclei there exists lifetimes of many years) and furthermore the energy of the system can be well-defined inside very narrow limits. Consequently one talks about well-defined excitation energies for nuclei and atoms and ascribes radioactive atoms and mesons certain masses. Still it is clear that such states cannot be stationary in the strict sense since—as the decay shows—it eventually changes in time and this shows again that the energy cannot be sharply defined. However, as we will see in the following, the minimal uncertainty in energy is inversely proportional to the lifetime, and consequently with increasing lifetime the state will become more and more like a stationary state*, and one speaks of quasi-stationary states. We will also consider one-dimensional models of decaying systems and investigate the question of lifetime and uncertainty in energy, partly through elementary considerations and partly through a more exact evaluation of a special example.

Let us consider a particle in the potential shown in Fig. 3.18. The energy spectrum is continuous so that all energies $E>0$ are possible. This is a consequence of the finite width of the barrier. If the barrier

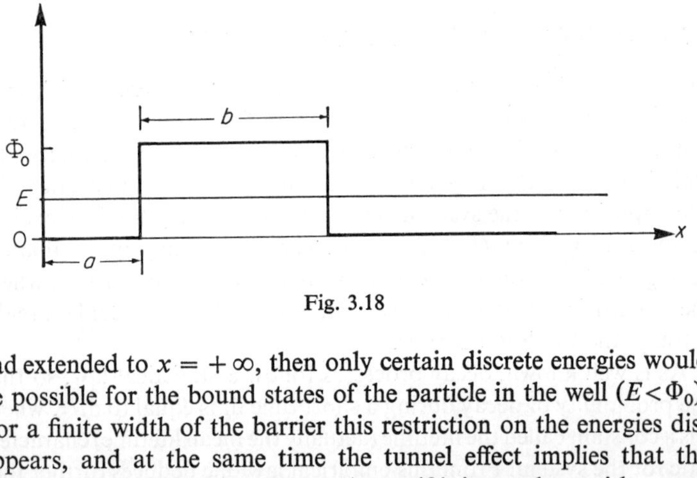

Fig. 3.18

had extended to $x = +\infty$, then only certain discrete energies would be possible for the bound states of the particle in the well ($E<\Phi_0$). For a finite width of the barrier this restriction on the energies disappears, and at the same time the tunnel effect implies that the particle does not remain in the well even if it is put there with energy less than Φ_0. However, when the transmission coefficient, T, of the barrier at the energy concerned is small, the particle may be regarded as being in a quasi-bound state. We can investigate the decay through a semiclassical reasoning. One imagines the motion in the well as an oscillation where the particle travels back and forth with the velocity $v = (2E/m)^{\frac{1}{2}}$ between the wall and the barrier. According to classical mechanics this is a complete description of the motion, but we will now take into consideration the small probability, T, of transmission

* This circumstance is utilized when one uses the convenient phrase that the decay consists of a transition from one stationary state to another.

every time the particle hits the barrier. In an interval of time dt this happens dt $v/2a$ times and the probability of the particle leaving the well during the time interval dt is therefore equal to dt $vT/2a$. This expression is in accordance with the empirical law of decay, and one sees that the mean lifetime, τ, for the bound state is given by

$$\frac{1}{\tau} = \frac{vT}{2a} = \sqrt{\frac{2E}{m}} \cdot \frac{T}{2a} = \nu_{cl} T \tag{3.7.1}$$

where we have introduced the classical frequency $\nu_{cl} = v/2a$ for the particle in the well. In this last form the expression for the lifetime can be used for a potential well and a barrier of arbitrary shape. Since $1/v_{cl}$ is a measure of the time the particle would spend in the well if there were no barrier, it is natural to demand that $T \ll 1$ before the state is called quasi-bound. It is noted that the above deduction of (3.7.1) is only justified if a is so large that it is possible to associate with the particle a wave packet which is narrow in comparison to the width of the well, and if at the same time T is constant for the energies within the wave packet. Although the deduction thus has a very limited applicability, the result is of general validity. We shall illustrate this later through an example where we avoid the wave packet description and study a quasi-stationary state with the smallest possible uncertainty in energy. This minimal uncertainty can also be evaluated by an elementary consideration. To see this we consider the situation of the system discussed above after the course of many lifetimes. The particle must then have left the well long ago and must now be travelling with velocity v along the x-axis. Since the particle was in the well at the time $t = 0$ and must have left it within a few lifetimes, it must now be on the x-axis in an interval of the order of magnitude $v\tau$. It must be described by a wave packet of similar dimensions and according to 3.1 one sees that the uncertainty in energy cannot be less than

$$\Delta E \approx v \Delta p \approx v \frac{h}{\Delta x} \approx \frac{h}{\tau}. \tag{3.7.2}$$

A similar argument can be carried through for any decay, and consequently the expression (3.7.2) always gives a lower limit for the uncertainty in energy of an unstable system. These fundamental facts are often formulated as a fourth uncertainty relation (Heisenberg's uncertainty relation for energy and time) and it can be interpreted as a fundamental impossibility of performing, during a period τ, a measurement of energy with an accuracy better than the one given by (3.7.2).

Until now we have only discussed lifetime and uncertainty of energy without examining why the quasi-stationary states have energies which lie close to certain discrete levels, such as are known, for instance, for α-radioactive nuclei. In order to understand this, one must have a closer look at the stationary wave functions. At the potential of Fig. 3.18 the stationary wave functions inside the barrier are of the form $A \exp(\beta x)+B \exp(-\beta x)$ when $E<\Phi_0$. The constants A and B are determined by the continuity conditions, and thus depend on the energy. For nearly all values of E, both A and B will be different from zero, and if the barrier is wide and high the wave function will be much larger on the outer side of the barrier than on the inner side. Only in the special case where $A \approx 0$ will the wave function decrease through the barrier. The two cases are illustrated in Fig. 3.19.

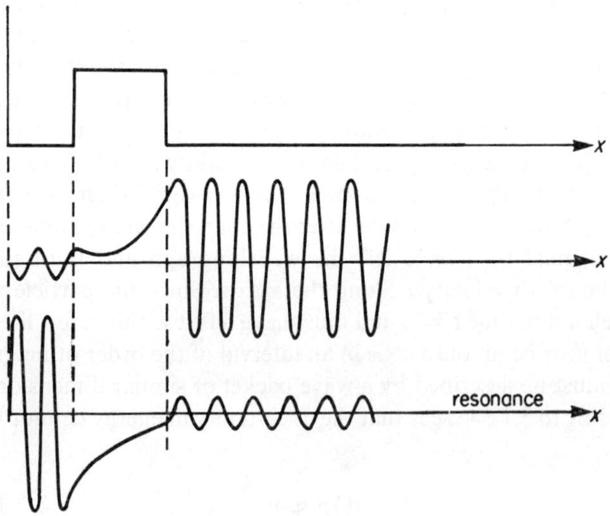

Fig. 3.19

The condition $A = 0$ is satisfied for just those energies which would form the discrete spectrum in the well if the barrier were infinitely wide. These energies are called resonance energies, and it is typical of resonance that as soon as the energy differs perceptibly from a resonance value the amplitude of the stationary wave functions becomes much larger outside the barrier than inside. Therefore, in order to build up a normalized wave function disappearing outside the barrier, one must make a superposition of stationary wave functions corresponding to energies close to the resonance values. From this it is seen that although all energies are possible for the potential considered, those states where we know with certainty that the particle has not left the well must have an energy distribution where the probability is concentrated on energies close to the resonance energies. We have thus defined a series of not quite sharp 'virtual levels'. A state where only energies from a single level enter may be termed quasi-stationary.

In the following we will, by explicit calculation for a simple example, determine the typical energy distribution for a quasi-stationary state. For this purpose we will use a summation of the wave function over the stationary wave functions which also allows us to determine the time-development of the state. The methods for this investigation were discussed in 2.5 where, however, we confined ourselves essentially to considering discrete energy spectra. Yet it is easy to generalize the results from 2.5 to include also continuous energy spectra by first enclosing the system in a potential box and afterwards letting the size of this tend towards infinity.

As long as the potential box has finite dimensions we can, according to 2.5, sum an arbitrary normalized wave function $\psi(x, 0)$ over normalized eigenstates $u_E(x)$. We thus have

$$\psi(x, 0) = \sum_n a_n u_{E_n}(x), \tag{3.7.3}$$

where

$$\int |u_{E_n}(x)|^2 \mathrm{d}x = 1. \tag{3.7.4}$$

The coefficients a_n are determined by the equation

$$a_n = \int \psi(x, 0)\, u^*_{E_n}(x) \mathrm{d}x. \tag{3.7.5}$$

The square of these quantities denotes the probability $P(E_n)$ of finding the energy E_n in the state described by ψ:

$$P(E_n) = |a_n|^2. \tag{3.7.6}$$

When the size of the potential box tends towards infinity the density of the energy states will increase so much as to make it reasonable to introduce a function $\mathrm{d}N/\mathrm{d}E$ indicating the number of states between E and $E+\mathrm{d}E$. We then find that the probability of finding the energy between E and $E+\mathrm{d}E$ is given by

$$P(E)\mathrm{d}E = |g(E)|^2\mathrm{d}E, \tag{3.7.7}$$

where

$$g(E) = a_n\sqrt{\frac{\mathrm{d}N}{\mathrm{d}E}}. \tag{3.7.8}$$

The function $g(E)$ is determined from (3.7.5) which, by a renormalization of u_{E_n}, can be written as

$$g(E) = \int \psi(x, 0)u^*(E, x)\mathrm{d}x, \tag{3.7.9}$$

where

$$u(E, x) = u_{E_n}(x)\sqrt{\frac{\mathrm{d}N}{\mathrm{d}E}}. \tag{3.7.10}$$

Finally, in the limit where the states lie infinitely close, it is possible to express also the series (3.7.3) by an integral. By means of (3.7.5) and (3.7.7) one finds that (3.7.3) can be written

$$\psi(x, 0) = \int g(E)u(E, x)\mathrm{d}E, \tag{3.7.11}$$

where one integrates over all energies.

In cases where one has a continuous spectrum in a certain energy interval and a discrete spectrum in another, (3.7.11) is naturally generalized to the result (2.5.17).

As a specific example, we will now investigate quasi-bound states in the potential shown in Fig. 3.20, where the calculations become especially simple.

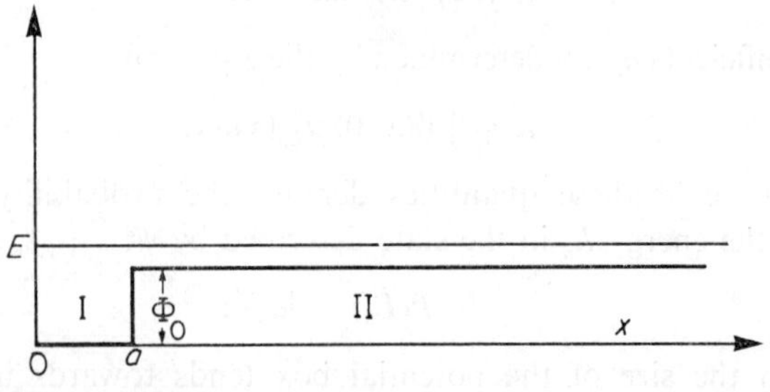

Fig. 3.20

For $E<\Phi_0$ the particle is tightly bound in the well and for $E-\Phi_0 \gg \Phi_0$ the particle escapes immediately. However, when E is only a little larger than Φ_0 the transmission coefficient at the potential barrier is small, and we then have the possibility of quasi-bound states.

Introducing the abbreviations

$$k = (2mE/\hbar^2)^{\frac{1}{2}} \quad \text{and} \quad k' = [2m(E-\Phi_0)/\hbar^2]^{\frac{1}{2}}, \tag{3.7.12}$$

we find the following acceptable solutions to the stationary Schrödinger equation:

$$u_{\text{I}}(x) = A \sin kx \tag{3.7.13}$$

$$u_{\text{II}}(x) = A\left[\sin ka \cos k'(x-a)+\frac{k}{k'}\cos ka \sin k'(x-a)\right], \tag{3.7.14}$$

where we have made use of the continuity conditions at $x = a$ and the fact that the wave function vanishes at $x = 0$. A is an arbitrary constant at our disposal for the normalization of the solutions.

If we consider the case $E-\Phi_0 \ll \Phi_0$, so that $k/k' \gg 1$, the amplitude of u_{II} will be much larger than that of u_{I}, unless $\cos ka \approx 0$. In this way certain k-values $k_n = (n+\frac{1}{2})\pi/a$ (where n is an integer) are distinguished, and according to (3.7.12) this corresponds to certain values of k'_n and energies E_n which appear as resonance energies. This is obvious when the function (3.7.14) is written in the form

$$u_{\text{II}} = B \sin(k'x+\delta) \tag{3.7.15}$$

since the amplitude B must then satisfy the equation

$$B^2 = A^2[\sin^2 ka+(k/k')^2 \cos^2 ka]. \tag{3.7.16}$$

The phase constant, δ, also depends on the energy, but at this point we shall not concern ourselves with the explicit expression for this quantity.

When the energy is close to a resonance value, E_n, so that $\sin^2 ka \approx 1$ and $\cos^2 ka \approx a^2(k-k_n)^2$, we may, to a good approximation, replace (3.7.16) by

$$B^2 = A^2[1+(k/k')^2a^2(k-k_n)^2],$$

or, using $k\mathrm{d}k = k'\mathrm{d}k'$,

$$B^2 = A^2\left[1+a^2\left(\frac{\mathrm{d}k'}{\mathrm{d}E}\right)_n^2 (E-E_n)^2\right] \text{(near resonance*)}. \tag{3.7.17}$$

The function u_{I} will, to a good approximation, be proportional to $\sin k_n x$ for all energies near resonance, and this is therefore the wave function inside the well for the quasi-stationary state at the virtual level in question. If we know with certainty at the time $t = 0$ that the particle is in the well, the normalized wave function must, at this time, be given approximately by

$$\psi(x, 0) = \begin{cases} \sqrt{2/a}\sin k_n x, & \text{for } x<a \\ 0 \quad , & \text{for } x>a, \end{cases} \tag{3.7.18}$$

as shown in Fig. 3.21.

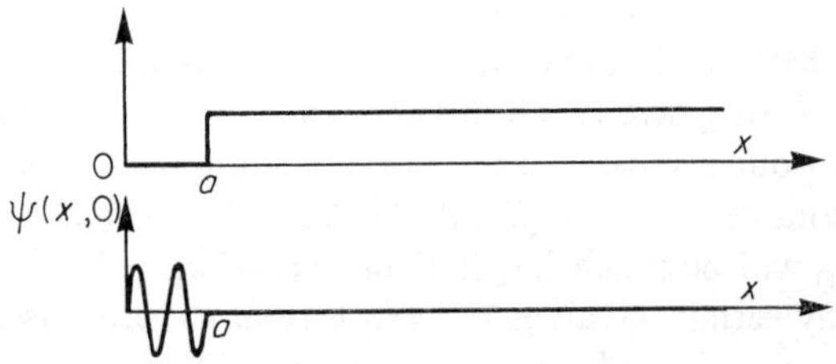

Fig. 3.21

We will sum this wave function over the stationary solutions near resonance, but for this purpose the functions $u(x)$ must be normalized. The introduction of a fixed wall at $x = L (L \gg a)$ makes the spectrum discrete since the requirement $u(L) = 0$ selects those solutions from (3.7.10) which satisfy

$$k'L+\delta = N\pi \quad \text{(where } N \text{ is an integer).} \tag{3.7.19}$$

The corresponding wave function can be normalized in the usual sense through the demand

$$\int_0^L |u_N(x)|^2 \mathrm{d}x = 1. \tag{3.7.20}$$

* This last transformation is permissible when $E-E_n \ll E-\Phi_0$.

When L is sufficiently large one may in (3.7.20) completely neglect the contribution from $x<a$ and one finds thus for $L\to\infty$

$$\int_0^L |u_N|^2 \mathrm{d}x \approx B^2 \int_a^L \sin^2(k'x+\delta)\mathrm{d}x \approx B^2 \frac{L}{2}. \tag{3.7.21}$$

Hence the stationary solutions in the box are normalized by putting $B = \sqrt{2/L}$; however, in order to obtain a normalization of eigenfunctions for the continuous spectrum we must furthermore take into account the density of states, $\mathrm{d}N/\mathrm{d}E$. When $L\to\infty$ one sees from (3.7.19) that the number of states in a wave interval $\mathrm{d}k'$ is equal to $L\mathrm{d}k'/\pi$, from which one obtains

$$\frac{\mathrm{d}N}{\mathrm{d}E} = \frac{L}{\pi}\frac{\mathrm{d}k'}{\mathrm{d}E}.$$

We therefore obtain normalized wave functions in the sense defined by (3.7.7) by choosing

$$B = \left(\frac{2}{\pi}\frac{\mathrm{d}k'}{\mathrm{d}E}\right)^{\frac{1}{2}}. \tag{3.7.22}$$

For energies close to the resonance where we can use (3.7.21), we thus find from (3.7.22) and (3.7.13) that the normalized stationary solutions for $x<a$ are, for $E\approx E_n$

$$u_{\mathrm{I}}(E, x) = \left(\frac{2}{\pi}\frac{\mathrm{d}k'}{\mathrm{d}E}\right)^{\frac{1}{2}} \frac{\sin kx}{[1+a^2(\mathrm{d}k'/\mathrm{d}E)_n{}^2(E-E_n)^2]^{\frac{1}{2}}}. \tag{3.7.23}$$

We can now expand the wave functions (3.7.18) on the normalized eigenfunctions for the continuous spectrum, so that the wave function is represented in the form (3.7.11) with $g(E)$ given by (3.7.9). In our case one finds

$$g(E) = \int_0^\infty u^*(E, x)\psi(x, 0)\mathrm{d}x = \int_0^a u_{\mathrm{I}}{}^*(E, x)\psi(x, 0)\mathrm{d}x, \tag{3.7.24}$$

where we have used the fact that $\psi(x, 0) = 0$ for $x>a$. Introducing the explicit expressions from (3.7.18) and (3.7.23), we obtain

$$g(E) = \left(\frac{4}{\pi a}\frac{\mathrm{d}k'}{\mathrm{d}E}\right)^{\frac{1}{2}} \frac{1}{[1+(\mathrm{d}k'/\mathrm{d}E)_n{}^2(E-E_n)^2a^2]^{\frac{1}{2}}} \int_0^a \sin kx \sin k_n x \,\mathrm{d}x. \tag{3.7.25}$$

When E is sufficiently close to E_n, the integral in (3.7.25) becomes equal to $a/2$, and we find the following approximation for $g(E)$:

$$g(E) = \frac{1}{\sqrt{\pi}} \frac{\Gamma_n/(2\Gamma)^{\frac{1}{2}}}{[(\Gamma_n/2)^2+(E-E_n)^2]^{\frac{1}{2}}}$$

$$\approx \frac{1}{\sqrt{\pi}} \frac{(\Gamma_n/2)^{\frac{1}{2}}}{[(\Gamma_n/2)^2+(E-E_n)^2]^{\frac{1}{2}}}, \tag{3.7.26}$$

where we have introduced the abbreviations

$$\Gamma = \frac{2}{a}\frac{\mathrm{d}E}{\mathrm{d}k'}, \qquad \Gamma_n = \frac{2}{a}\left(\frac{\mathrm{d}E}{\mathrm{d}k'}\right)_n. \tag{3.7.27}$$

The probability distribution of the energy is, according to (3.7.7) and (3.7.26), determined by

$$P(E) = |g(E)|^2 \approx \frac{1}{\pi}\frac{\Gamma_n/2}{(\Gamma_n/2)^2+(E-E_n)^2}. \tag{3.7.28}$$

One sees that $P(E)$ is reduced by a factor of two when the energy changes from E_n to $E_n \pm \Gamma_n/2$ so that Γ_n indicates the width at half-maximum in the energy distribution (see Fig. 3.22). Because of this property the energy Γ_n is denoted the width of the virtual level.

Now we are also able to follow in detail how the quasi-stationary state (3.7.18) develops in time. Since time variation of the stationary wave functions is given by the phase factor $\exp(-\mathrm{i}Et/\hbar)$, one finds at the time t [see (2.5.18)]:

$$\psi(x, t) = \int g(E)u(E, x) \exp(-\mathrm{i}Et/\hbar)\mathrm{d}E. \tag{3.7.29}$$

For $x<a$ one thus obtains from (3.7.23, 26, 27):

$$\psi(x, t) \approx \frac{1}{\pi}\sqrt{\frac{2}{a}} \int_0^\infty \frac{\Gamma_n/2 \exp(-\mathrm{i}Et/\hbar)}{(\Gamma_n/2)^2+(E-E_n)^2} \sin kx \,\mathrm{d}E$$

$$\approx \sqrt{\frac{2}{a}} \exp(-\Gamma_n t/2\hbar) \sin k_n x \exp(-\mathrm{i}E_n t/\hbar). \tag{3.7.30}$$

The probability density in the well is therefore equal to

$$\frac{2}{a} \sin^2 k_n x \exp(-\Gamma_n t/\hbar),$$

and one finds the following total probability for the particle to be in the well at the time t:

$$P(t) = \int_0^a |\psi(x, t)|^2 = \exp(-\Gamma_n t/\hbar). \qquad (3.7.31)$$

This expression is exactly what was to be expected according to the empirical law of decay, and it is seen that the mean lifetime is given by

$$\tau = \frac{\hbar}{\Gamma_n}, \qquad (3.7.32)$$

which corresponds to (3.7.2) when Γ_n is taken as a measure of the uncertainty in energy.

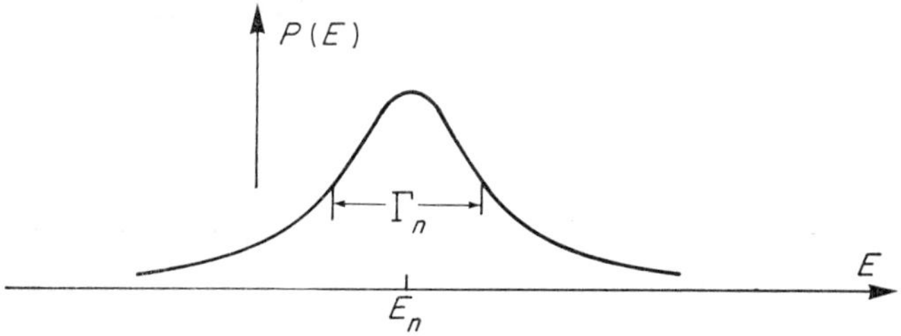

Fig. 3.22

Here we have only studied the change in time of the wave function inside the well. A corresponding discussion of the wave-function outside the well will be somewhat more complicated because one must take into consideration the dependence of the phase δ, in (3.7.15), on the energy. We will only indicate the result in Fig. 3.23 by sketching the probability density at various times.

One sees how the probability density leaks out of the well. In the region $x > a$, where the wave function is zero at the beginning, a complex wave function is formed in the course of time, corresponding to a purely outgoing current. The distribution of the probability density outside the well can be understood qualitatively when one considers the fact that the particle cannot have left the well before the time $t = 0$, and that the probability of decay during a certain interval of time decreases steadily.

The expressions (3.7.28) and (3.7.32), giving the connexion between energy distribution and lifetime, are typical for quasi-stationary states. One has only a good description of these states if it is possible to define the resonance energy and lifetime and, in this connexion, it is interesting to see that for the latter purpose one can use the simple deduction leading to (3.7.1). It is easy to verify this relation for the example discussed above. The transmission coefficient of the potential jump is, according to (3.3.4), given by $T = 4k'/k$ for $k' \ll k$, from which one gets, by applying (3.7.1): $1/\tau = 2\hbar k'/ma$, which is again identical to what one finds from (3.7.32), (3.7.27) and (3.7.12). A comparison of the expressions (3.7.1) and (3.7.32) may serve to estimate how sharply the virtual levels will stand out. An oscillating wave packet in the well must in fact be formed by the superposition of states from several virtual levels; the classical frequency is then determined by $\nu_{cl} = \Delta E/h$ (cf. 2.5), where ΔE denotes the distance between the virtual levels. According to (3.7.1) the reciprocal lifetime

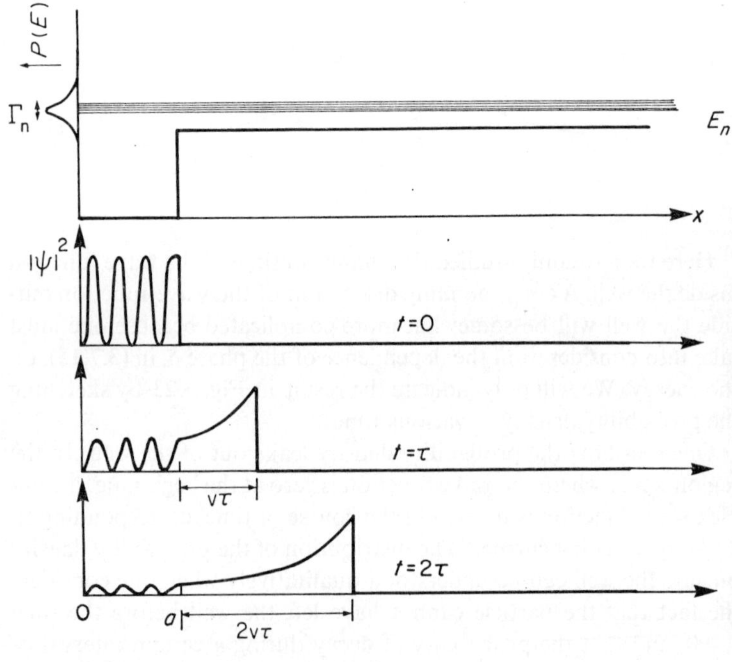

Fig. 3.23

is equal to $T\Delta E/h$ which by comparison with (3.7.32) gives

$$\frac{\Gamma}{\Delta E} = \frac{T}{2\pi}.$$

It is thus seen that one must demand that $T \ll 1$ in order that the levels be well separated, which is a condition for the validity of the notion of quasi-stationary states. For potentials, as in the case of those considered here, the transmission coefficient increases with increasing energy. Thus the lowest virtual levels will be the narrowest ones, and the levels will overlap completely when the energy becomes so large that quasi-bound states no longer exist.

Until now we have only considered systems with one possibility of decay. In practice there is often a possibility of decay through several channels as, for instance, by the potential shown in Fig. 3.24, where

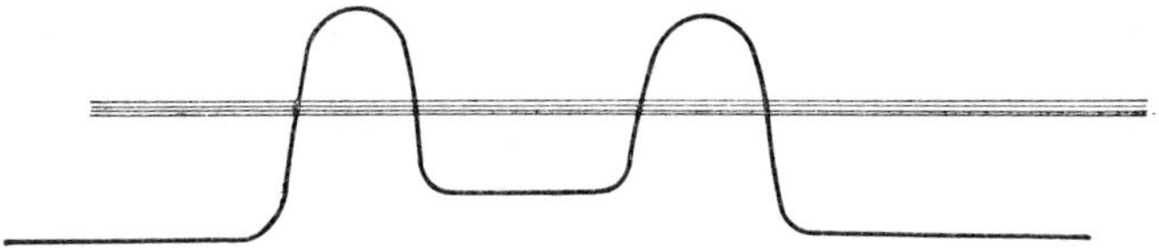

Fig. 3.24

a virtual level is indicated. The lifetime is here determined by the coefficients of transmission T_1 and T_2 for both potential barriers, and apparently one finds

$$\frac{1}{\tau} = \nu_{cl}(T_1 + T_2). \tag{3.7.33}$$

In accordance with this the width of the level, Γ, is equal to the sum of the partial widths, $\Gamma_1 = \nu_{cl}T_1$ and $\Gamma_2 = \nu_{cl}T_2$, indicating the probability of decay through the two channels. Note that lifetime and energy spread belong to the level and not to the individual decay modes. Consequently one measures Γ and τ by studying decays through one of the channels only. By counting the decays through both channels it is furthermore possible to determine the branching ratio, Γ_1/Γ_2.

Problems

3.7.1 Investigate the lowest virtual levels in a potential well like the one in Fig. 3.18. Assume that $E \ll \Phi_0$ and $\kappa = mb^2\Phi_0/2\hbar^2 / \gg 1$, and determine how Γ and ΔE vary with level number. Carry out an investigation of the normalized stationary solutions near resonance.

3.7.2 Find a complete expression for the wave function (3.7.10) near resonance and perform the normalization. Use then the expression (3.7.26), together with the weight factor (3.7.29) in order to determine the evolution in time of the wave function (3.7.18) in the region $x>a$. How is the wave function affected if one observes at the time t that decay has not yet taken place?

3.7.3 Calculate, for the potential well in Fig. 3.20, the values of E/Φ_0 and Γ_n for the lowest virtual levels. As example set

$$\kappa = ma^2\Phi_0/2\hbar^2 = 36.$$

3.7.4 Interpret the wave functions in Fig. 3.19 as a stationary description of the reflexion of a particle arriving from the right. Discuss the qualitative difference in the reflexion when the energy of the particle is respectively close to, or far away from, a resonance.

3.7.5 A particle is in a stationary bound state with energy E in the potential shown on the figure (full drawn). At the time $t = 0$ the potential changes suddenly to the dotted curve. It is assumed that the change is made in such

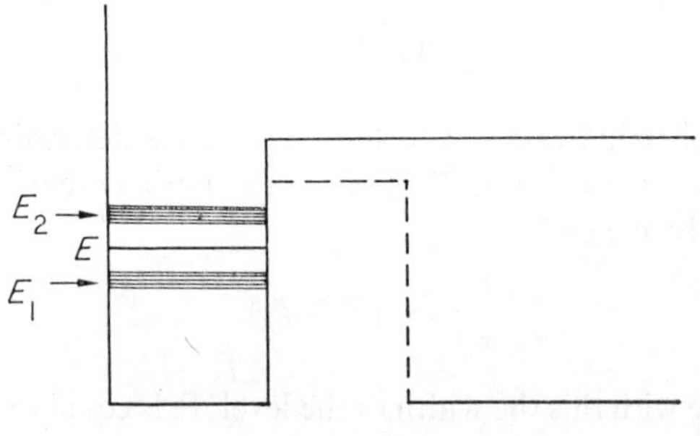

Fig. 3.25

a way that immediately afterwards the particle finds itself in a superposition of states corresponding to two well-separated virtual levels at E_1 and E_2 with widths Γ_1 and Γ_2. The energy is distributed with the probabilities P_1 and P_2 on the two levels.* Discuss the result of an experimental investigation of the decay process, partly if one registers particles as a function of time only, and partly when one also measures their energy.

* These probabilities are also called the populations of the levels.

3.8 TIME-DEPENDENT PERTURBATION THEORY AND THE GOLDEN RULE

The exponential decay of a quasi-stationary state is a basic phenomenon in atomic and nuclear physics. We have already studied the decay from a state in a potential well where the particle slowly leaks out through a potential barrier. An apparently different type of approach is the time-dependent perturbation theory, which we shall treat below. The lifetime of a state is given here by a celebrated formula, the so-called Golden Rule. It may perhaps seem as if the exponential decay is a more or less accidental result. Actually it is a rather fundamental and simple property of quantum states, and could be obtained from formulae like (2.1.1–4).

The problem to be studied now is that of an initial state, rather well-defined with respect to energy, and for which transitions are possible to a continuum set of states. We might, for instance, think of an excited atom which decays to the ground state plus an emitted photon, as discussed qualitatively in 4.6.

In order to study such examples we use at first an exact time-dependent formulation. Consider an unperturbed Hamilton operator $\hat{H}_0$, for which the eigenfunctions $\psi_n{}^0 = v_n(\mathbf{r}) \exp(-iE_n t/\hbar)$ are known, i.e. $\hat{H}_0 \psi_n{}^0 = E_n \psi_n{}^0$. The total Hamilton operator is $\hat{H} = \hat{H}_0 + \hat{\Omega}$, so that the wave function ψ of the system must satisfy the basic Schrödinger equation

$$(\hat{H}_0 + \hat{\Omega})\psi = i\hbar\dot{\psi}. \tag{3.8.1}$$

In this equation we may expand the wave function ψ over the complete set of states $\psi_n{}^0$,

$$\psi = \sum_n a_n(t)\psi_n{}^0, \tag{3.8.2}$$

where the amplitudes a_n are expected to depend on time. When this expression is introduced in the Schrödinger equation (3.8.1), we can solve with respect to $\dot{a}_n$ (in fact, multiply by ψ_n^{0*} in (3.8.1) and integrate over space, using the orthonormality of the $\psi_n{}^0$) and an alternative formulation of (3.8.1) is obtained.

$$i\hbar\dot{a}_n(t) = \int d^3r \psi_n^{0*}(\mathbf{r}, t)\hat{\Omega}\psi(\mathbf{r}, t). \tag{3.8.3}$$

It may seem strange that the formulation (3.8.3) is of advantage, since the desired change of a_n is expressed in terms of the unknown total wave function ψ. The point is, however, that crude but well-chosen guesses of ψ in (3.8.3) give good first estimates of $\dot{a}_n$.

We may rewrite formula (3.8.3) in terms of matrix elements of the perturbation $\hat{\Omega}$.

$$\mathrm{i}\hbar\dot{a}_n(t) = \sum_m a_m(t)\Omega_{nm} \exp\left[-\frac{\mathrm{i}}{\hbar}(E_m - E_n)t\right], \tag{3.8.4}$$

where

$$\Omega_{nm} = \int \mathrm{d}^3 r v_n^*(\mathbf{r})\hat{\Omega}v_m(\mathbf{r}).$$

The perturbation connects the original state, n_0 (e.g. the excited atom), with a set of final states, n (the atom in the ground state plus a light quantum, which may be defined by its energy and direction of motion). Since the space available to the outgoing particle, or light quantum, is unrestricted, the final set is a continuum set of states, so that all energies are possible. The perturbation $\hat{\Omega}$ thus allows transitions from n_0 to n, whereas these states would remain separate and stationary with the original Hamilton operator $\hat{H}_0$. We have not yet explicitly stated anything about a possible time-dependence of $\hat{\Omega}$, but from now on we assume that $\hat{\Omega}$ is independent of t.

We are only interested in the transitions $n_0 \rightleftharpoons n$. A general perturbation $\hat{\Omega}$ could have many other effects. Thus if $\Omega_{n_0n_0} \neq 0$, this merely means that the energy level n_0 is shifted by $\Omega_{n_0n_0}$, as is also seen in the stationary perturbation treatment, 3.6; similarly the Ω_{nn} are not of direct interest. It is also reasonable to disregard the matrix elements $\Omega_{n'n''}$ between final states. The effect of such extra transitions is expected to be negligible, one reason being that they only give rise to terms involving higher powers of Ω, which is assumed to be a small perturbation. The only matrix elements of interest are Ω_{n_0n}, and we simply assume that all other matrix elements are zero.

Equation (3.8.4) may then be written

$$\mathrm{i}\hbar\dot{a}_n(t) = a_{n_0}(t)\Omega_{nn_0} \exp\left[-\frac{\mathrm{i}}{\hbar}(E_{n_0} - E_n)t\right], \tag{3.8.5a}$$

$$\mathrm{i}\hbar\dot{a}_{n_0}(t) = \sum_n a_n(t)\Omega_{n_0n} \exp\left[-\frac{\mathrm{i}}{\hbar}(E_n - E_{n_0})t\right], \tag{3.8.5b}$$

where the states labelled n are all different from the state n_0. The reader may easily verify that, like (3.8.2), the set of equations (3.8.5) conserves the total probability, i.e.

$$|a_{n_0}(t)|^2+\sum_n |a_n(t)|^2 = \text{const.} = 1, \tag{3.8.6}$$

as was to be expected.

The simplest way of solving (3.8.5) is to suppose that the state n_0 decays exponentially in time, so that we make the trial solution

$$a_{n_0}(t) = \exp(-\lambda t), \tag{3.8.7}$$

where we have assumed that $a_{n_0}(0) = 1$, and therefore $a_n(0) = 0$. The lifetime, $\tau = 1/2\lambda$ is as yet unknown, and we shall see that it can be determined in a self-consistent manner.

With the ansatz (3.8.7), the equations for a_n in (3.8.5) may be integrated. We find

$$a_n(t) = -\Omega_{nn_0}\frac{\exp[-\lambda t+\mathrm{i}(E_n-E_{n_0})t/\hbar]-1}{E_n-E_{n_0}+\mathrm{i}\hbar\lambda}. \tag{3.8.8}$$

One method, at this stage, would be to insert (3.8.8) in (3.8.5b), whereby the a_n's are eliminated, and a consistent condition is obtained. We shall use an alternative, and slightly simpler method. We eliminate the a_n's by introducing (3.8.8) in (3.8.6). Therefore we first form the numerical square of (3.8.8):

$$\sum_n |a_n(t)|^2 = \sum_n |\Omega_{nn_0}|^2\frac{|\exp[-\lambda t+\mathrm{i}(E_n-E_{n_0})t/\hbar]-1|^2}{(E_n-E_{n_0})^2+\hbar^2\lambda^2}. \tag{3.8.9}$$

We use now the assumption that the final states n are exceedingly closely spaced, and replace $\sum$ by $\int \mathrm{d}E\rho(E)$, where $\rho(E)$ is the density of final states per unit energy interval. Moreover, the denominator in (3.8.9) is small only in a narrow interval of energies around E_{n_0}, of width $\hbar\lambda$. We can therefore normally assume that $\rho(E)$ is a constant equal to $\rho(E_{n_0})$ and take it outside the integral. Finally, we can suppose that for the narrow region of states in question the matrix elements $|\Omega_{nn_0}|^2$ are also constant and can be taken outside the integral. The simplified form of (3.8.9) is therefore

$$\sum_n |a_n(t)|^2 = \rho(E_{n_0})|\Omega_{nn_0}|^2\int_{-\infty}^{+\infty}\mathrm{d}E\,\frac{|\exp[-\lambda t+\mathrm{i}(E_n-E_{n_0})t/\hbar]-1|^2}{(E-E_{n_0})^2+\hbar^2\lambda^2}, \tag{3.8.10.}$$

where we have put $\pm\infty$ on the limits in order to indicate that the integration extends far on both sides of E_{n_0}. The definite integral is easily calculated (cf., for example, Pierce, *Table of Integrals*, Ciros and Co., Boston), and becomes $[1-\exp(-2\lambda t)]\pi/\lambda\hbar$. We insert this result on the right-hand side of (3.8.10). On the left-hand side, we express $\sum |a_n(t)|^2$ by means of (3.8.6) and (3.8.7) as a function of the unknown parameter λ, and thus obtain

$$1-\exp(-2\lambda t) = \frac{\pi}{\lambda\hbar}|\Omega_{nn_0}|^2\rho(E_{n_0})\,[1-\exp(-2\lambda t)]. \quad (3.8.11)$$

Therefore the assumption of exponential decay in (3.8.7) is consistent, and (3.8.11) becomes an identity if

$$\frac{1}{\tau} = 2\lambda = \frac{2\pi}{\hbar}|\Omega_{nn_0}|^2\rho(E_{n_0}). \quad (3.8.12)$$

This formula is the previously mentioned Golden Rule, by means of which numerous types of decay phenomena can be calculated.

and

$$\frac{1}{\sin\theta}\frac{\mathrm{d}}{\mathrm{d}\theta}\left(\sin\theta\frac{\mathrm{d}\boldsymbol{\theta}}{\mathrm{d}\theta}\right)+\left(\lambda-\frac{\nu}{\sin^2\theta}\right)\boldsymbol{\theta}=0. \tag{4.1.10}$$

The first of these equations is seen at once to have solutions of the type $\exp(\pm\mathrm{i}\phi\sqrt{\nu})$. This function changes by a factor $\exp(\pm\mathrm{i}\sqrt{\nu}2\pi)$ when ϕ increases from 0 to 2π. Thus, in order to have a one-to-one correspondence between the wave function and the position of the particle, one must demand that $\sqrt{\nu}$ be equal to an integer, so that

$$\nu = m^2, \quad m = 0, \pm 1, \pm 2, \pm 3, \ldots \;. \tag{4.1.11}$$

If one lets m take on both negative and positive values the acceptable solutions to (4.1.9) may be written in the form

$$\boldsymbol{\phi}_m = \frac{1}{\sqrt{2\pi}}\exp(\mathrm{i}m\phi), \tag{4.1.12}$$

where the normalization factor $1/\sqrt{2\pi}$ is chosen such that

$$\int_0^{2\pi}|\boldsymbol{\phi}_m|^2\mathrm{d}\phi = 1.$$

In the equation for $\boldsymbol{\theta}$, (4.1.10), we must now introduce $\nu = m^2$, from which it is seen that the solutions to (4.1.10) only depend on m through the numerical value $|m|$. It is convenient to introduce the substitutions

$$\cos\theta = \mu, \qquad \boldsymbol{\theta}(\theta) = P(\mu), \tag{4.1.13}$$

whereby (4.1.10) is transformed into

$$(1-\mu^2)\frac{\mathrm{d}^2P}{\mathrm{d}\mu^2}-2\mu\frac{\mathrm{d}P}{\mathrm{d}\mu}+\left(\lambda-\frac{m^2}{1-\mu^2}\right)P=0. \tag{4.1.14}$$

This is the so-called Legendre differential equation. The solutions may be investigated through a method similar to the one used in 2.4 in connexion with the harmonic oscillator. We will not go through this discussion here* but only describe the results. Generally, the

---

* Introducing $P = (1-\mu^2)^{|m|/2}f(\mu)$ one may solve the equation (4.1.14) by inserting a power series for f. In order that $P(\mu)$ be finite in the entire interval $-1\leqslant\mu\leqslant 1$, one finds that this series must break off after a finite number of terms.

solutions of (4.1.14) tend towards infinity for $\mu \to \pm 1$. Solutions which remain finite in the entire interval $-1 \leqslant \mu \leqslant 1$, only exist if

$$\lambda = l(l+1), \qquad l = 0, 1, 2, \cdots . \tag{4.1.15}$$

In the special case $m = 0$, the solutions are the so-called Legendre polynomials which can be written in the form

$$P_l(\mu) = \frac{1}{2^l l!} \frac{\mathrm{d}^l}{\mathrm{d}\mu^l} (\mu^2 - 1)^l, \tag{4.1.16}$$

from which it is evident that $P_l(\mu)$ is a polynomial of degree l in μ. The Legendre polynomials are also known from the theory of electromagnetism, where they appear in the multipole expansion of electromagnetic potentials.

For $m \neq 0$ the acceptable solutions to (4.1.14) [with $\lambda = l(l+1)$] are the associated Legendre functions $P_l^{|m|}(\mu)$ which can be derived from the Legendre polynomials (4.1.16) through the relation

$$P_l^{|m|}(\mu) = (1-\mu^2)^{\frac{|m|}{2}} \frac{\mathrm{d}^{|m|}}{\mathrm{d}\mu^{|m|}} P_l(\mu), \tag{4.1.17}$$

so that in particular $P_l^0(\mu) = P_l(\mu)$.

Moreover one sees from (4.1.17), using that $P_l(\mu)$ is a polynomial of lth degree in μ, that $P_l^{|m|}(\mu)$ vanishes identically when $m > l$. Acceptable solutions to (4.1.14) thus only exist when λ is of the form (4.1.15) and when furthermore $-l \leqslant m \leqslant l$.

We may summarize the discussion above by saying that (4.1.7) has acceptable solutions only when λ satisfies (4.1.15), and for a given value of l the solutions may be written as

$$\begin{aligned} Y_{l,m}(\theta, \phi) &= N P_l^{|m|} (\cos \theta) \exp (im\phi), \\ m &= -l, -l+1, \ldots, l-1, l, \end{aligned} \tag{4.1.18}$$

where N is a normalization constant. Apparently, for a given value of l, (4.1.7) has $2l+1$ independent acceptable solutions corresponding to the choices $m = -l, -l+1, \ldots . l$. The numbers l and m identifying the states are called the aximuthal and the magnetic quantum numbers respectively. The functions $Y_{l\,m}(\theta, \phi)$ given by (4.1.18) determine, according to (4.1.5), the angular dependence of the wave function in a state with quantum numbers l and m. The functions are called the spherical harmonics. They form an orthogonal system such that the integral $\int Y_{l,m}^* Y_{l',m'} d\Omega$ over all solid angles is zero unless $l = l'$ and $m = m'$. The spherical functions can be normalized by choosing the proportionality factor N in (4.1.18) to be

$$N = \left(\frac{2l+1}{4\pi} \frac{(l-|m|)!}{(l+|m|)!} \right)^{\frac{1}{2}} \times \begin{cases} 1 \text{ for } m<0 \\ (-1)^m \text{ for } m>0 \end{cases} \tag{4.1.19}$$

so that

$$\int_0^{2\pi} \int_0^{\pi} Y_{l,m}^* Y_{l',m'} \sin\theta \, d\theta \, d\phi = \delta_{l,l'} \delta_{m,m'}. \tag{4.1.20}$$

The orthogonal system formed by the spherical functions is complete, so that the angular dependence of an arbitrary wave function can be represented by a superposition of spherical harmonics. Explicit expressions for the spherical harmonics can be calculated from (4.1.13), (4.1.16) and (4.1.17). With the normalization constant (4.1.19), one finds for $l = 0$ and 1

$$Y_{1,1} = -\sqrt{\frac{3}{8\pi}} \sin\theta \exp(i\phi),$$

$$Y_{0,0} = \frac{1}{\sqrt{4\pi}}, \qquad Y_{1,0} = \sqrt{\frac{3}{4\pi}} \cos\theta,$$

$$Y_{1,-1} = \sqrt{\frac{3}{8\pi}} \sin\theta \exp(-i\phi). \tag{4.1.21}$$

We conclude this section with a few remarks as to how the above results affect the final determination of the stationary states, which are obtained by the solution of the radial equation (4.1.6). In this equation one must insert $\lambda = l(l+1)$, and for a given value of l one will in general find acceptable solutions for a number of discrete energy values (if the particle is in a bound state). The solutions to the radial equation can be numbered with a quantum number n_r (the radial quantum number) which, together with l and m, gives a complete characterization of the state. According to a notation which is taken over from atomic spectroscopy, the states are labelled by the letters s, p, d, f, g, h, and so on, in alphabetical order, according to whether $l = 0, 1, 2, 3, 4, 5, \cdots$. Since the magnetic quantum number m does not appear in the radial equation, one sees that an energy level belonging to a state with given l must be at least $2l+1$ times degenerate*. We will later show that if we add a weak magnetic field to the central field, the energy levels for a charged particle split up according to m. Hence the name for this quantum number.

Finally, it should be mentioned that by the solution of the radial equation it may be useful to introduce the substitution

$$R(r) = \frac{\chi(r)}{r}, \tag{4.1.22}$$

which when inserted in (4.1.6) [with $\lambda = l(l+1)$] gives

$$-\frac{\hbar^2}{2m}\frac{d^2\chi}{dr^2}+\left(\Phi(r)+\frac{\hbar^2 l(l+1)}{2mr^2}\right)\chi = E\chi. \tag{4.1.23}$$

This equation is formally identical with the stationary Schrödinger equation for a particle in a one-dimensional potential

$$\Phi_{\text{eff}}(r) = \Phi(r)+\frac{\hbar^2 l(l+1)}{2mr^2},$$

i.e. the sum of the potential and the so-called centrifugal potential. Using this analogy one must remember that r only runs from 0 to $+\infty$, and that a boundary condition: $R(r)$ finite for $r\to 0$, implies $\chi(r)\to 0$ for $r\to 0$. These features may be introduced in the one-dimensional analogy by adding an infinitely high potential wall at $r = 0$.

Problems

4.1.1 Introduce the substitution $P(\mu) = (1-\mu^2)^{\frac{|m|}{2}} f(\mu)$ in (4.1.14), and show by induction the validity of (4.1.17).

4.1.2 Find, from (4.1.16), an explicit expression for the coefficients of the individual powers of μ in the polynomial $P_l(\mu)$ and show that $P_l(\mu)$ is a solution of (4.1.14) with $\lambda = l(l+1)$, $m = 0$.

4.1.3 Show, by the help of (4.1.16), that the following formula applies for the Legendre polynomials:

$$\int_{-1}^{1} P_l(\mu)P_{l'}(\mu)d\mu = \frac{2}{2l+1}\,\delta_{l,l'}.$$

4.1.4 Find the number of nodes for $P_l^{|m|}(\mu)$ in the open interval -1 to 1. Apparently these zeros correspond to node-surfaces of the wave function

* A higher degree of degeneracy may appear if by solving the radial equation one finds that several l-values belong to the same energy. We will see an example of this later.

$R(r)\,Y_{l,m}(\theta, \phi)$. The main quantum number n is often introduced as $1+$ the total number of node surfaces in the real part of the wave function. Indicate the different types of node surfaces for the above-mentioned wave function and find, for a fixed value of l and m, the smallest value of n.

4.1.5 Show that for a given value of l the radial equation (4.1.6) has only one solution for every possible value of E. Show that solutions belonging to different values of E are orthogonal so that

$$\int_0^\infty R^*_{E,l}(r)R_{E',l}(r)r^2\mathrm{d}r = 0 \text{ for } E \neq E'.$$

4.1.6 Show that the solutions found by separation of (4.1.1) cover all possibilities so that any stationary wave function may be written as a linear combination of solutions $R_l(r)\,Y_{l,m}(\theta, \phi)$, where $R_l(r)$ satisfies (4.1.6) with $\lambda = l(l+1)$. Utilize the completeness of the spherical harmonics.

4.2 ANGULAR MOMENTUM

In the classical description of central motion it is convenient to utilize the fact that the angular momentum is constant in time. Thus the orbit of a particle in a conservative central potential can be characterized by the magnitude of the total energy and the magnitude and direction of the angular momentum. According to 3.4, it follows from the theorem of Ehrenfest that also in quantum mechanics the angular momentum vector is a constant of the motion. In the present section we shall look into this matter and show that the quantum numbers l and m introduced in 4.1 characterize the magnitude of the angular momentum and its direction with respect to the polar axis.

A physical interpretation of the quantum numbers can be obtained in a quite elementary way. It is well known from classical mechanics of a particle that the radial motion in a central field can be treated as a one-dimensional motion about a fixed axis if one adds a potential accounting for the centrifugal force. This force points outwards and is of magnitude $m_0 r\omega^2$ where ω is the instantaneous angular velocity in the orbit. The constant angular momentum has the magnitude $|\mathbf{L}| = m_0 r^2\omega$, and the centrifugal force may thus be derived from a potential $L^2/(2m_0 r^2)$, which one must add to the potential of the central field $\Phi(r)$ to achieve a one-dimensional treatment of the radial motion. In 4.1 we saw how a quantum mechanical description of the radial motion could be treated as a one-dimensional motion (4.1.23) through the addition of a potential

$\hbar^2 l(l+1)/(2m_0 r^2)$, and this is seen to be identical to the classical centrifugal potential if one identifies the two quantities L^2 and $\hbar^2 l(l+1)$.

On these grounds it is plausible to assume that the stationary states, discussed in 4.1 with quantum numbers l, must be associated with an angular momentum of the size

$$|\mathbf{L}| = \sqrt{L^2} = (L_x{}^2+L_y{}^2+L_z{}^2)^{\frac{1}{2}} = \hbar[l(l+1)]^{\frac{1}{2}}.$$

However, in order to have a precise definition of angular momentum, we must introduce the operators associated with angular momentum. With the help of these it is possible to calculate expectation values of arbitrary powers of the components of angular momentum, and thus determine the probability distribution of angular momentum in a state in exactly the same way as we have earlier discussed the determination of the energy distribution from the energy operator $\hat{H}$.

The angular momentum operators can be constructed from the previously introduced momentum operators. First we will see, however, how it is possible in a convenient way to find certain angular momentum operators in polar coordinates by comparing the Hamiltonian used in (4.1.4),

$$\hat{H} = -\frac{\hbar^2}{2m_0}\left\{\frac{1}{r}\frac{\partial^2}{\partial r^2}r + \frac{1}{r^2}\left[\frac{1}{\sin\theta}\frac{\partial}{\partial\theta}\sin\theta\frac{\partial}{\partial\theta} + \frac{1}{\sin^2\theta}\frac{\partial^2}{\partial\phi^2}\right]\right\}+\Phi(r), \tag{4.2.1}$$

with the classical Hamilton function in polar coordinates.

The kinetic energy is given by

$$T = \tfrac{1}{2}m_0(\dot{r}^2+r^2\dot{\theta}^2+r^2\sin^2\theta\,\dot{\phi}^2),$$

from which generalized momenta are found to be

$$\begin{aligned} p_r &= m_0\dot{r}, \\ p_\theta &= m_0 r^2\dot{\theta}, \\ p_\phi &= m_0 r^2\sin^2\theta\dot{\phi}. \end{aligned} \tag{4.2.2}$$

When one eliminates the generalized velocities in the expression for T and adds the potential energy, one obtains the Hamilton function

$$H = \tfrac{1}{2}m_0\left[p_r{}^2+\frac{1}{r^2}\left(p_\theta{}^2+\frac{1}{\sin^2\theta}p_\phi{}^2\right)\right]+\Phi(r). \tag{4.2.3}$$

We saw formerly in 1.6 that in Cartesian coordinates the operator $\hat{H}$ is derived from the classical Hamilton function by replacing the classical momenta by the corresponding differential operators. We assume that the same connection between H and $\hat{H}$ can be achieved with generalized coordinates. By a comparison of (4.2.1) and (4.2.3) we can then find the differential operators which must be associated with the squares of the momenta p_r^2, p_θ^2 and p_ϕ^2. Apparently, of these three quantities the last two are concerned with the angular momentum. The exact meaning is easily seen from (4.2.2) when one evaluates the contribution to $\mathbf{L}$ from the r, θ and ϕ components of the velocity. One finds that p_ϕ is the projection of the angular momentum, L_z, on the z-axis, and that L^2 is given by

$$L^2 = p_\theta{}^2 + \frac{1}{\sin^2\theta} p_\phi{}^2.$$

By comparison between (4.2.1) and (4.2.3) we thus find the operator assignment

$$\hat{L}^2 = -\hbar^2 \left\{ \frac{1}{\sin\theta}\frac{\partial}{\partial\theta}\sin\theta\frac{\partial}{\partial\theta} + \frac{1}{\sin^2\theta}\frac{\partial^2}{\partial\phi^2} \right\} \tag{4.2.4}$$

and

$$\hat{L}_z{}^2 = -\hbar^2 \frac{\partial}{\partial\phi^2} \tag{4.2.5}$$

and we may write

$$\hat{H} = -\frac{\hbar^2}{2m}\frac{1}{r}\frac{\partial^2}{\partial r^2} r + \frac{\hat{L}^2}{2mr^2} + \Phi(r). \tag{4.2.6}$$

The expressions (4.2.4) and (4.2.5) for the operators can also be obtained in a more systematic—but more tedious—way from the operators for the linear momentum $\hat{p}_x = -i\hbar\partial/\partial x$, etc. The operators for the angular momentum are found by substituting $\mathbf{p}$ with $\hat{\mathbf{p}}$ in the classical expression for the angular momentum $\mathbf{L} = \mathbf{r}\times\mathbf{p}$; thus

$$\begin{aligned} \hat{L}_x &= y\hat{p}_z - z\hat{p}_y = -i\hbar\left(y\frac{\partial}{\partial z} - z\frac{\partial}{\partial y}\right) \\ \hat{L}_y &= z\hat{p}_x - x\hat{p}_z = -i\hbar\left(z\frac{\partial}{\partial x} - x\frac{\partial}{\partial z}\right) \\ \hat{L}_z &= x\hat{p}_y - y\hat{p}_x = -i\hbar\left(x\frac{\partial}{\partial y} - y\frac{\partial}{\partial x}\right). \end{aligned} \tag{4.2.7}$$

Introducing the polar coordinates (4.1.2) it is possible to show that

$$\begin{aligned}\hat{L}_x &= i\hbar\left(\sin\phi\frac{\partial}{\partial\theta}+\cot\theta\cos\phi\frac{\partial}{\partial\phi}\right)\\ \hat{L}_y &= i\hbar\left(-\cos\phi\frac{\partial}{\partial\theta}+\cot\theta\sin\phi\frac{\partial}{\partial\phi}\right)\\ \hat{L}_z &= -i\hbar\frac{\partial}{\partial\phi}.\end{aligned}\tag{4.2.8}$$

Evaluating the expressions $\hat{L}^2 = \hat{L}_x{}^2+\hat{L}_y{}^2+\hat{L}_z{}^2$ and $\hat{L}_z{}^2$, this is seen to be in accordance with (4.2.4) and (4.2.5).

We see now that the angular equation (4.1.7), discussed in 4.1, can be written by means of (4.2.4) in the form

$$\hat{L}^2 Y(\theta, \phi) = \hbar^2\lambda Y(\theta, \phi). \tag{4.2.9}$$

The solutions to this equation, namely the spherical harmonics $Y_{l,m}$, are thus eigenfunctions of the operator $\hat{L}^2$ corresponding to the eigenvalue $\hbar^2 l(l+1)$. We conclude that in a state where the angular dependence of the wave function is given by a spherical harmonic $Y_{l,m}(\theta, \phi)$, the square of the angular momentum of the particle has the definite value $\hbar^2 l(l+1)$. From the expression (4.1.18) it is seen immediately that $Y_{l,m}$ is also an eigenfunction of the operator $\hat{L}_z$, with eigenvalue $\hbar m$, and consequently L_z has the definite value $\hbar m$.

The results found are in complete accordance with the discussion in 3.4. Beforehand, we would expect that the vector of angular momentum was a constant of motion and it is also possible by means of the expressions (4.2.4)–(4.2.8) to show that $\hat{\mathbf{L}}$ commutes with $\hat{H}$, i.e.

$$[\hat{\mathbf{L}}, \hat{H}] = 0. \tag{4.2.10}$$

This equation also shows that it is possible to find states being at the same time eigenstates for both $\hat{H}$ and either $\hat{L}_x$, $\hat{L}_y$ or $\hat{L}_z$. In order to find states in which both L_x, L_y and L_z have definite values one must demand, however, that the operators $\hat{L}_x$, $\hat{L}_y$ and $\hat{L}_z$ also commute mutually. Yet this is not the case. Using the expressions (4.2.7) or (4.2.8) one finds the following important commutation relations:

$$\begin{aligned}[][\hat{L}_x, \hat{L}_y] &= i\hbar\hat{L}_z,\\ [\hat{L}_y, \hat{L}_z] &= i\hbar\hat{L}_x,\\ [\hat{L}_z, \hat{L}_x] &= i\hbar\hat{L}_y.\end{aligned}\tag{4.2.11}$$

The fact that $\hat{L}_x$, $\hat{L}_y$ and $\hat{L}_z$ are non-commuting, while the three operators $\hat{p}_x$, $\hat{p}_y$ and $\hat{p}_z$ commute, causes the quantum mechanical description of angular momenta to be far more complicated than the description of linear momenta. In general we cannot find states with more than one component of angular momentum having a definite value, and, as above, we must confine ourselves to specifying, for instance, L_z. However, a direct consequence of the commutation relations (4.2.11) is that even though $\hat{L}_x$, $\hat{L}_y$ and $\hat{L}_z$ do not commute mutually, they all commute with $\hat{L}^2$, i.e.

$$[\hat{L}^2, \hat{\mathbf{L}}] = 0. \tag{4.2.12}$$

Furthermore, since $\hat{L}^2$ commutes with $\hat{H}$ [see, for instance, (4.2.6)]:

$$[\hat{L}^2, \hat{H}] = 0, \tag{4.2.13}$$

it is possible to find common eigenfunctions for, say, $\hat{H}$ and $\hat{L}_z$, which are at the same time eigenfunctions for $\hat{L}^2$. The set of eigenfunctions u_{nlm} found in 4.1 is just a set with these properties. It satisfies simultaneously the three eigenvalue equations

$$\begin{aligned} \hat{H}u_{nlm} &= E_n u_{nlm}, \\ \hat{L}^2 u_{nlm} &= \hbar^2 l(l+1) u_{nlm}, \\ \hat{L}_z u_{nlm} &= \hbar m u_{nlm}. \end{aligned} \tag{4.2.14}$$

Usually the solutions to (4.2.14) are, as we saw, unambiguously determined. One says that with $\hat{H}$, $\hat{L}^2$ and $\hat{L}_z$ one has found a complete set of commuting constants of the motion. Of course, this does not prevent the existence of many other constants of the motion. As examples of this we have already seen $\hat{L}_x$ and $\hat{L}_y$. As another example we mention the parity. It follows directly from the expression (4.1.1) for $\hat{H}$ that the parity operator $\hat{P}$ is a constant of the motion [see (3.4.9)]. The operator $\hat{P}$ also commutes with the operators of angular momentum. This is connected with the fact that $\mathbf{L}$ is an axial vector, whose orientation depends upon whether the coordinate system is a right-hand or a left-hand system. It also follows directly from the expressions (4.2.7) that

$$[\hat{P}, \mathbf{L}] = 0, \tag{4.2.15}$$

and consequently

$$[\hat{P}, \hat{L}^2] = 0. \tag{4.2.16}$$

Thus the operator $\hat{P}$ has eigenfunctions in common with $\hat{H}$, $\hat{L}^2$ and

$\hat{L}_z$, and since the solutions u_{nlm} of (4.2.14) are unambiguously determined, the function u_{nlm} is already a common eigenfunction. The eigenvalues for $\hat{P}$ can be determined directly from the expressions (4.1.5), (4.1.17) and (4.1.18) for u_{nlm}, using the definition of $\hat{P}$ in polar coordinates

$$\hat{P}f(r, \theta, \phi) = f(r, \pi-\theta, \phi+\pi). \tag{4.2.17}$$

One finds

$$\hat{P}u_{nlm} = (-1)^l u_{nlm} \tag{4.2.18}$$

which shows that the parity is $(-1)^l$.

We can represent the angular momentum by a vector diagram as shown in Fig. 4.1 for the state $l = 3$. The magnitude of the angular momentum is in this case $\sqrt{3(3+1)}\,\hbar = 3{\cdot}47\,\hbar$, and it can be oriented in 7 directions with respect to the z-axis. The angular part of the wave

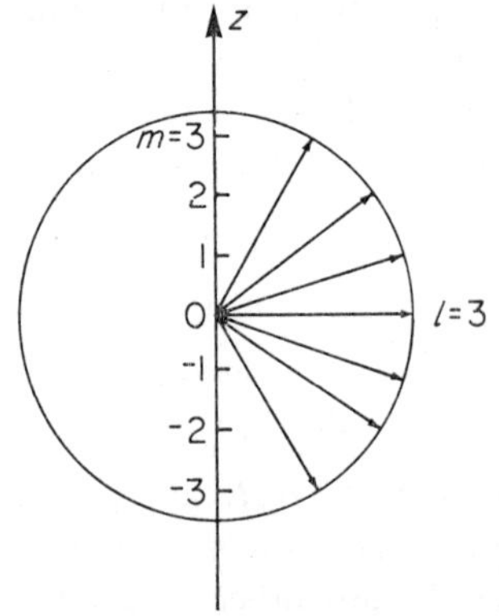

Fig. 4.1

function is in the 7 cases given by the spherical harmonics $Y_{3,-3}$, $Y_{3,-2}, \cdots\cdots, Y_{3,3}$. The quantum condition $|m| \leqslant l$ mentioned in 4.1 has the obvious significance that the projection of the angular momentum on a line cannot exceed its length. It is also seen that the degeneracy of the energy with respect to m is a natural consequence of the fact that the energy in a central potential is independent of the direction of the angular momentum.

Since we cannot ascribe definite values to L_x and L_y in the states considered above, it is apparently meaningless to draw the angular momentum vector in a definite position with respect to the x- and y-axes on the vector diagram in Fig. 4.1. By means of the operators

(4.2.7) we can calculate the expectation value $\langle L_x \rangle$ and $\langle L_y \rangle$, and it is easily seen that both of these are equal to zero when the angular part of the wave function is of the form (4.1.18). Thus in the state considered we must imagine that the angular momentum vector has a given length and a given direction with respect to the z-axis; consequently it also has a definite length of projection on the xy-plane, but the direction of this last component of the angular momentum is completely undetermined. One also arrives at the same picture by considering the angular distribution of the probability density. Since $|Y_{lm}|^2$ is independent of ϕ it is seen that the probability density is axially symmetric about the z-axis and it is thus not possible to even approximately define the intersection between the plane of the orbit and the xy-plane. On the other hand, it is possible to speak about a certain slope of the orbital plane, which is expressed most clearly in the probability density at large values of l. When, for instance, $m = l$, the spherical harmonics is proportional to $\sin^l \theta$, which for $l \gg 1$ only deviates essentially from zero for $\theta \approx \pi/2$, i.e. when the particle is close to the xy-plane.

In the above we have chosen the z-axis as polar axis, which especially distinguished the angular momentum component L_z. However, it is obvious from the physical symmetry between all three axes that we can also find states where, for example, L_x has a definite value. In this case L_z must be undetermined, and thus the angular dependence of the wave function cannot be given by one of the spherical harmonics. Since these however, form a complete orthogonal system, we can write the angular dependence of the wave function as a superposition of the functions $Y_{l,m}$, and the absolute square of the coefficients of the expansion must then be interpreted as the probabilities that L^2 and L_z have the values belonging to $Y_{l,m}$ in the new state. [See problem 4.2.3.]

Finally, we will study how the stationary states for a charged particle change when a weak magnetic field is added to the pure central potential which we have been considering until now. In order to find the corresponding Hamilton operator, we will base our investigations on the classical Hamiltonian for a particle in a magnetic field, which, for a particle with mass m_0 and charge q, is given by

$$H = \frac{1}{2m_0}\left[\mathbf{p} - \frac{q}{c}\mathbf{A}(\mathbf{r})\right]^2 + \Phi(r), \tag{4.2.19}$$

where $\mathbf{A}(\mathbf{r})$ denotes the vector potential due to the magnetic field and $\Phi(r)$ is the potential energy from the non-magnetic forces. It is seen that the magnetic field is taken into consideration by substituting $\mathbf{p}-q\mathbf{A}/c$ for $\mathbf{p}$ in H_0 (the Hamiltonian without magnetic field). For sufficiently weak magnetic fields we can discard terms of second order in $\mathbf{A}$, and thus obtain the approximate Hamiltonian

$$H = \frac{1}{2m_0}\left(|\mathbf{p}|^2 - 2\frac{q}{c}\,\mathbf{p}\cdot\mathbf{A}\right) + \Phi(r) = H_0 - \frac{q}{m_0 c}\,\mathbf{p}\cdot\mathbf{A}. \tag{4.2.20}$$

We now assume the magnetic field $\mathbf{B}$ to be homogenous and oriented along the z-axis, i.e. $\mathbf{B} = (0, 0, B)$. This field is to be derived from the vector potential through the relation $\mathbf{B} = \text{curl}\,\mathbf{A}$, which can be achieved by introducing $\mathbf{A} = (-y, x, 0)B/2$. The scalar product $\mathbf{p}\cdot\mathbf{A}$ in (4.2.20) then becomes

$$\mathbf{p}\cdot\mathbf{A} = \frac{B}{2}(-yp_x + xp_y) = \frac{B}{2}L_z,$$

and consequently the Hamiltonian may be written

$$H = H_0 - \frac{qB}{2m_0 c}L_z.$$

The Hamilton operator is thus given by

$$\hat{H} = \hat{H}_0 - \frac{qB}{2m_0 c}\hat{L}_z, \tag{4.2.21}$$

where $\hat{H}_0$ is the Hamilton operator for the central field.

If the solutions for the pure central potential are known, it is now extremely simple to indicate the solutions to the stationary Schrödinger equation with the Hamiltonian (4.2.21). Corresponding to a level with energy E_0 in the pure central potential there are $2l+1$ eigenfunctions for $\hat{L}_z$ with eigenvalues $m\hbar$. These wave functions are apparently also eigenfunctions for the Hamiltonian (4.2.21), since we find

$$\begin{aligned} \hat{H}u_{E_0,l,m} &= \left(\hat{H}_0 - \frac{Bq}{2m_0 c}\hat{L}_z\right)u_{E_0,l,m} \\ &= \left(E_0 - \frac{Bq}{2m_0 c}\hbar m\right)u_{E_0,l,m}. \end{aligned} \tag{4.2.22}$$

If the stationary wave functions for the pure central potential are chosen in the above-mentioned manner, they are thus automatically solutions also of the stationary Schrödinger equation with a weak magnetic field. It follows from the completeness of the system of eigenfunctions that we need not look for additional solutions.

The stationary wave functions for the central field can thus be carried over unchanged; however, the energy will now depend on the quantum number m. The $2l+1$ stationary states, which all had the energy E_0 in the central field, will now correspond to $2l+1$ equidistant energy levels. This splitting is called the Zeeman effect.

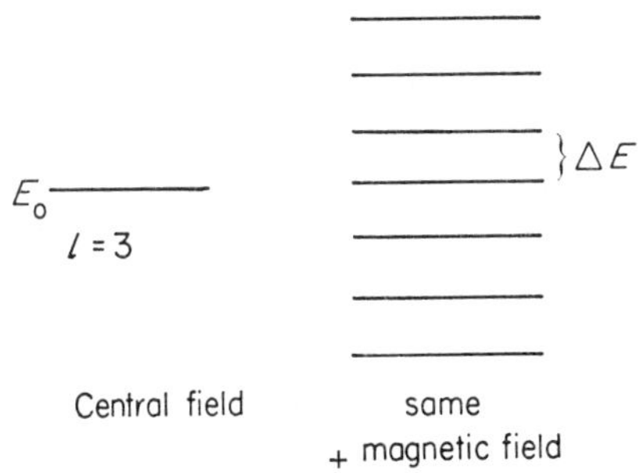

Fig. 4.2

The distance between two neighbouring levels in such a series is seen from (4.2.22) to be $\Delta E = \hbar Bq/2m_0c$ (compare Fig. 4.2). In order to obtain a physical picture of the stationary states in the magnetic field one may perform a superposition of wave functions with the same E_0 and l but different values of m, and it is noted that one then obtains a probability density which depends periodically on the time through ϕ [compare problem 4.2.7]. The angular velocity corresponding to this periodic motion is $\omega_{cl} = \Delta E/\hbar = Bq/2m_0c$, which is the Larmor frequency for a particle with charge q and mass m_0 in the field B. This frequency is known to be of importance also in the classical description of the motion of a charged particle in a central potential plus a homogenous magnetic field. For the pure central field, the classical orbit lies in a fixed plane. With an additional magnetic field, the angular momentum of the particle and the plane of the orbit precess around the direction of the magnetic field. The angular velocity of precession is just equal to the Larmor frequency.

We thus have the simple picture that the introduction of a magnetic field simply gives rise to a precession of the angular momentum vector and splitting of the energy levels. The fact that the precession does not show up in the probability densities of the stationary states means that the phase of the precession is completely undetermined. The energy difference between states with different magnetic quantum number, m, can be understood in a simple way by noting that the orbiting charged particle represents a magnetic dipole. The magnetic moment $\mathbf{M}$ of this dipole is given by $\mathbf{M} = \mathbf{L}q/(2m_0c)$, where $\mathbf{L}$ is the angular momentum of the particle*. The energy of the dipole in the magnetic field is then $-\mathbf{M}\cdot\mathbf{B} = -L_zqB/(2m_0c) = -qm\hbar B/(2m_0c)$, in agreement with (4.2.22).

This picture of precession can be supported by considering the motion of the components of the angular momentum. It is thus seen from (4.2.13) that L^2 and L_z are constants of the motion, as in the pure central field ($\hat{L}^2$ and $\hat{L}_z$ both commute with $\hat{H}_0$ and $\hat{L}_z$ and consequently also with $\hat{H}$). In the magnetic field, L_x and L_y are not constants of the motion, since their operators do not commute with $\hat{L}_z$. For the expectation values of L_x and L_y one can easily show [problem 4.2.7] that

$$\frac{\mathrm{d}}{\mathrm{d}t}\langle L_x\rangle = -\frac{qB}{2m_0c}\langle L_y\rangle$$

and

$$\frac{\mathrm{d}}{\mathrm{d}t}\langle L_y\rangle = \frac{qB}{2m_0c}\langle L_x\rangle.$$

The expectation value of the angular momentum thus precesses in exactly the same way as the classical angular momentum, in accordance with the theorem of Ehrenfest.

Problems

4.2.1 Find, for $l \gg 1$, approximate expressions for the magnitude of the angular momentum and for the minimum angle with the z-axis in an eigenstate for L^2 and L_z.

---

* The natural unit for the magnetic moment in atomic physics is the so-called Bohr magneton $\mu_B = e\hbar/2m_ec$, which derives from this formula by setting $L = \hbar$, while q and m_0 are put equal to the electron charge e and mass m_e respectively.

4.2.2 Show that in an eigenstate for L^2 and L_z the following relation holds for the fluctuations σ_{L_x} and σ_{L_y}:

$$\sigma_{L_x}^2 \sigma_{L_y}^2 = \hbar^2[l(l+1) - m^2],$$

and find, for given l, the minimum value of this quantity.

4.2.3 Show that a normalized wave function $\psi(r, \theta, \phi)$ can always be expressed by the series

$$\psi = \sum_{l,m} f_{lm}(r) Y_{lm}(\theta, \phi)$$

and that the quantity

$$|C_{lm}|^2 = \int_0^\infty |f_{lm}(r)|^2 r^2 \mathrm{d}r$$

must be interpreted as the probability for L^2 and L_z to have the values $\hbar^2 l(l+1)$ and $\hbar m$ respectively.

4.2.4 Consider a *p*-state where L_x has exactly the value 0. Find the probability distribution for L_z.

4.2.5 For a three-dimensional harmonic oscillator with the potential $\frac{1}{2}kr^2$, the stationary wave functions can be written in the form $u_x(x)u_y(z)u_z(z)$, where the three functions u_x, u_y and u_z are stationary solutions for one-dimensional oscillators on the *x*-, *y*- and *z*-axes, respectively. Consider the ground state of the three-dimensional oscillator and the three states of the next level which correspond to the excitation of the *x*, *y* and *z* oscillators, respectively. Find, for all four states, the probability distribution of L^2 and L_y.

4.2.6 Consider a state of a charged particle in a central potential where E and L have definite values while L_z with equal probability has the values $m\hbar$ and $(m+1)\hbar$. How does the probability density change in time? At a certain moment one adds a homogenous magnetic field parallel to the *z*-axis. How does the probability distribution now vary in time?

4.2.7 For a charged particle in a central field plus a weak homogenous magnetic field along the *z*-axis, compute the time derivatives

$$\frac{\mathrm{d}}{\mathrm{d}t}\langle L_x\rangle, \frac{\mathrm{d}}{\mathrm{d}t}\langle L_y\rangle \text{ and } \frac{\mathrm{d}}{\mathrm{d}t}\langle L_z\rangle,$$

and show that the expectation value of the angular momentum vector precesses around the *z*-axis.

4.3 COULOMB POTENTIAL

In this and the following paragraphs we will investigate the stationary states for a particle in some special central fields which are of importance in atomic and nuclear physics. We will base the discussion on the results of 4.1 and we will thus look for solutions to the stationary Schrödinger equation of the form

$$u(r, \theta, \phi) = R(r) Y_{l,m}(\theta, \phi). \tag{4.3.1}$$

For the cases in question, we then merely have to solve the radial equation (4.1.6) for each value of the azimuthal quantum number $l = 0, 1, 2, \cdots$. By introducing

$$R(r) = \frac{\chi(r)}{r}, \tag{4.3.2}$$

the radial equation may be written in the form (4.1.23), i.e.

$$-\frac{\hbar^2}{2m}\chi'' + \left[\Phi(r) + \frac{\hbar^2 l(l+1)}{2m_0 r^2}\right]\chi = E\chi. \tag{4.3.3}$$

As mentioned in 4.1, the determination of the possible eigenvalues is hereby formally reduced to a one-dimensional problem for a particle on the r-axis in the effective potential

$$\Phi_{\text{eff}} = \Phi(r) + \frac{\hbar^2 l(l+1)}{2m_0 r^2}.$$

From the analogy with the one-dimensional problems treated earlier, it is seen that the energy spectrum will be continuous for $E > \Phi(\infty)$. When the potential is attractive there will furthermore be a possibility for bound states, with a discrete energy spectrum, in the interval $(\Phi_{\text{eff}})_{\min} < E < \Phi(\infty)$. In the following we are particularly interested in these bound states.

In this section we consider the Coulomb field around a positive charge of magnitude Ze (a nucleus). The potential energy in this field for a particle with charge $-e$ (e.g. an electron) is then $\Phi(r) = -Ze^2/r$. The corresponding effective potential for the radial equation is shown in Fig. 4.3 for s-, p- and d-states. It is seen that the bound states must have a negative total energy and thus for these states we may write $E = -|E|$. The radial equation (4.3.3) may then be written

$$\frac{d^2\chi}{dr^2} + \left(-\frac{2m_0|E|}{\hbar^2} + \frac{2m_0 Ze^2}{\hbar^2}\frac{1}{r} - \frac{l(l+1)}{r^2}\right)\chi = 0, \tag{4.3.4}$$

which apparently has the asymptotic solutions

For $r \to 0$,	for $r \to \infty$.
$\chi \sim r^{l+1}$,	$\chi \sim \exp\left[-\frac{r}{\hbar}(2m\lvert E\rvert)^{\frac{1}{2}}\right]$
$\chi \sim r^{-l}$,	$\chi \sim \exp\left[\frac{r}{\hbar}(2m\lvert E\rvert)^{\frac{1}{2}}\right]$

Of these solutions only the upper set leads to acceptable wave functions, since $l \geqslant 0$ and $\chi(r)$ must tend towards zero for $r \to 0$ and ∞.

It is convenient to introduce the dimensionless variable

$$\xi = (2m_0\lvert E\rvert)^{\frac{1}{2}}\frac{2r}{\hbar} \tag{4.3.5}$$

which, substituted in (4.3.4), gives

$$\chi''(\xi)+\left(-\frac{1}{4}+\frac{\lambda}{\xi}-\frac{l(l+1)}{\xi^2}\right)\chi(\xi) = 0 \tag{4.3.6}$$

with

$$E = \frac{m_0 Z^2 e^4}{2\hbar^2\lambda^2}. \tag{4.3.7}$$

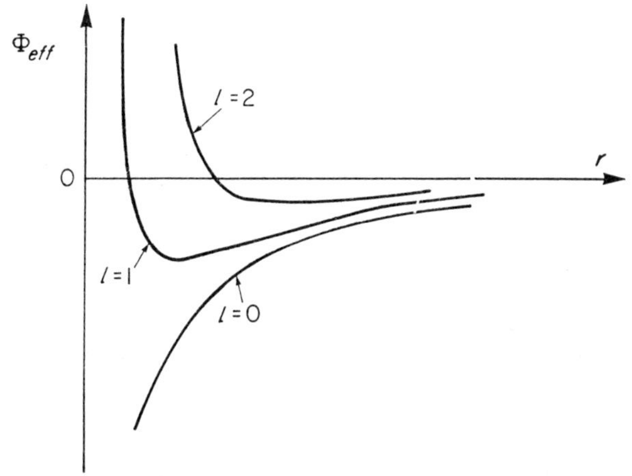

Fig. 4.3

The asymptotic solutions are now ξ^{l+1} and $\exp(-\xi/2)$, and it is thus natural to write χ in the form

$$\chi(\xi) = \xi^{l+1} \exp(-\xi/2)\, L(\xi), \tag{4.3.8}$$

which, inserted in (4.3.6), leads to the differential equation

$$\xi L'' + [2(l+1) - \xi]\, L' + (\lambda - l - 1)\, L = 0. \tag{4.3.9}$$

We now look for solutions of this equation which tend towards a constant when $\xi \to 0$ and which increase more slowly than $\exp(\xi/2)$ when $\xi \to \infty$. Such solutions exist only for a discrete set of λ-values and the solutions are polynomials which can be identified (apart from a constant factor) with the so-called generalized Laguerre polynomials.

It is easy to find the possible values of λ and a recurrence formula for the coefficients in the polynomial solutions by writing the solutions of (4.3.9) as a power series:

$$L(\xi) = \sum_{\nu=0}^{\infty} a_\nu \xi^\nu. \tag{4.3.10}$$

Introducing this series into (4.3.9) one obtains the recurrence formula

$$\frac{a_{\nu+1}}{a_\nu} = \frac{\nu + l + 1 - \lambda}{(\nu+1)(\nu+2l+2)}. \tag{4.3.11}$$

For large values of ν one finds $a_{\nu+1}/a_\nu \approx 1/\nu$, which is the recurrence formula for the series

$$\sum_{\nu=0}^{\infty} \frac{1}{\nu!} \xi^\nu.$$

This series represents the function $\exp(\xi)$, and consequently the solutions are only acceptable when the series (4.3.10) terminates. According to (4.3.11), λ must then be an integer, and since furthermore $\nu \geqslant 0$, one obtains for a given l the following possible λ-values:

$$\lambda = n, \quad n = l+1, l+2, \cdots \quad . \tag{4.3.12}$$

Thus for given values of n and l the degree of the polynomial $L(\xi)$ is

$$\nu_{\max} = n - l - 1.$$

The number $n - l$ is also denoted by n_r and is called the radial quantum number, whereas n is called the main quantum number. The radial quantum number minus one indicates the number of zeros in the radial wave function. It indicates a numbering of the energy levels for a given l.

From (4.3.12) and (4.3.7) we find the following energy spectrum for the bound states*

$$E_n = -|E| = -\frac{m_0 Z^2 e^4}{2\hbar^2}\frac{1}{n^2}, \quad n = 1, 2, \cdots \tag{4.3.13}$$

The discrete spectrum is seen to contain infinitely many levels, the lowest of which are indicated in Fig. 4.4, together with the continuous spectrum for $E>0$. For a given value of l the quantum number n in (4.3.13) is limited by $n \geqslant l+1$, but apart from this the energy only depends on l and n_r through the combination $n_r + l = n$. This strong degeneracy is characteristic of the Coulomb field. In general the levels in a central field depend on both l and n_r or—stated in another way—on l and n. The energy levels are then often denoted by the main, or principal, quantum number followed by a letter indicating the l-value.

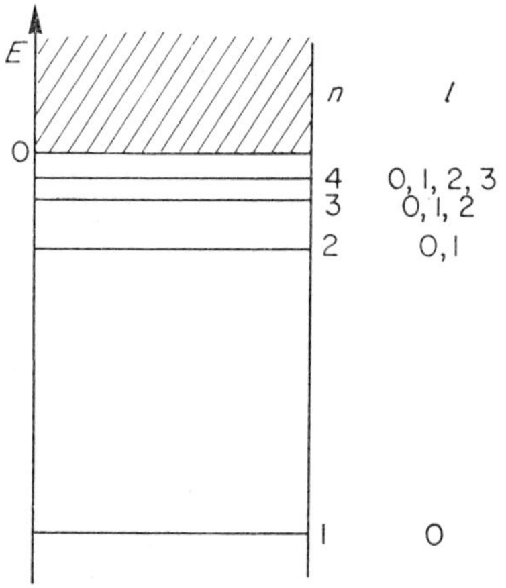

Fig. 4.4

With this notation the ground state in Fig. 4.4 is denoted by 1*s*. The second level may be regarded as a coincidence of the 2*s* and 2*p* states, the third level is coincidence of 3*s*, 3*p* and 3*d*, etc.

---

* When m_0 is equal to the mass of the electron m_e the quantity $m_e e^4/(2\hbar^2)$ is Rydberg's constant, which is equal to 13·6 eV.

The nth level in the Coulomb field is seen to correspond to states with azimuthal quantum numbers $l = 0, 1, 2, \ldots, n-1$, and since each of these have a degeneracy $2l+1$, the total degree of degeneracy is

$$\sum_{l=0}^{n-1} (2l+1) = \tfrac{1}{2}n\,(1+2n-1) = n^2.$$

Let us discuss briefly the wave functions corresponding to the bound states in the Coulomb field. Using (4.3.5) and (4.3.13) one finds

$$\xi = \frac{2Zr}{a_0 n}; \quad a_0 = \frac{\hbar^2}{m_0 e^2}, \tag{4.3.14}$$

where the quantity a_0 is the so-called Bohr radius for the hydrogen atom (when m_0 is the mass of the electron*). According to (4.3.2) and (4.3.8) the radial part of the wave function has the form

$$R_{n,l}(r) = N\xi^l \exp(-\xi/2)\,L(\xi), \tag{4.3.15}$$

where ξ is given by (4.3.14) and $L(\xi)$ is a polynomial of degree $n-l-1$. N is a normalization constant chosen in such a way that

$$\int |R_{n,l}(r)\,Y_{,lm}(\theta, \phi)\,|^2 d\tau = 1,$$

or

$$\int_0^\infty |R_{n,l}(r)|^2 r^2 dr = 1.$$

The polynomials $L(\xi)$ of low degree are easily calculated from the recurrence formula (4.3.11), with $\lambda = n$. After normalization one finds

$$\begin{aligned} 1s : R_{1,0} &= \left(\frac{Z}{a_0}\right)^{\frac{3}{2}} 2 \exp[-Zr/a_0] \\ 2s : R_{2,0} &= \left(\frac{Z}{2a_0}\right)^{\frac{3}{2}} \left(2-\frac{Zr}{a_0}\right) \exp(-Zr/2a_0] \\ 2p : R_{2,1} &= \left(\frac{Z}{2a_0}\right)^{\frac{3}{2}} \frac{Zr}{a_0\sqrt{3}} \exp(-Zr/2a_0]. \end{aligned} \tag{4.3.16}$$

From the general expression (4.3.15) it is seen that the wave function (and thus also the probability density) is non-vanishing for $r = 0$ only

---

* In this case one has $a_0 = 0{\cdot}529 \times 10^{-8}$ cm.

for s-states. The probability density behaves asymptotically as r^{2l} for $r \to 0$ so that the particle 'avoids' the origin more and more with increasing values of l. This is an effect of the centrifugal potential, and may be understood from the classical analogy where we may take l as a measure of the magnitude of the angular momentum. For s-states (angular momentum zero) the classical elliptical orbit degenerates into a straight line through the centre of force corresponding to the finite probability density for s-states at the centre. However, when comparing with the classical picture, one must remember that the angular part of the wave function in an s-state is given by the spherical harmonics $Y_{0\,0}$, which is a constant. Therefore the probability density is spherically symmetric and the direction of the 'linear orbit' is completely undefined.

Instead of the probability density it is often useful to consider the radial distribution function $D(r) = r^2|R(r)|^2$, which is defined such that $D(r)\mathrm{d}r$ indicates the probability of finding the particle in distances between r and $r+\mathrm{d}r$. Fig. 4.5 shows the radial wave function $R(r)$

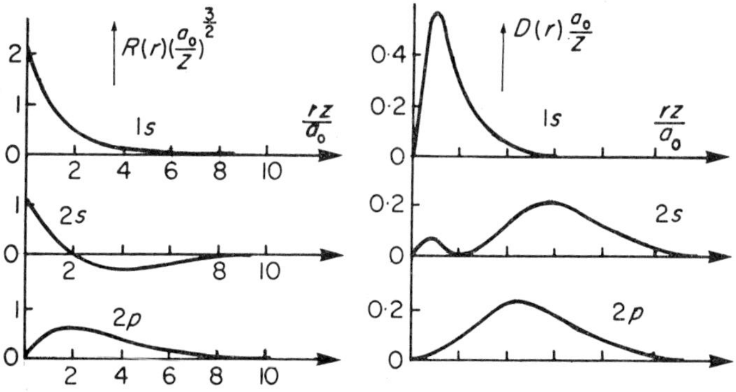

Fig. 4.5

and distribution $D(r)$ for the states (4.3.16). In the ground state it is easy to see from the explicit expression for $D(r)$ that $D(r)$ has a maximum at $r = a_0/Z$, which is thus the most probable distance for the particles. Similarly, one finds the mean distance $\langle r \rangle = 3a_0/(2Z)$. The general expression for the mean distance may be shown [by means of the formula in problem 4.3.6] to be

$$\langle r \rangle_{n,l} = \frac{n^2 a_0}{Z}\left[1+\tfrac{1}{2}\left(1-\frac{l(l+1)}{n^2}\right)\right], \tag{4.3.17}$$

from which it is seen that the quantity $n^2 a_0/Z$ may be used as a crude measure for the dimension of the system.

Finally, if we want to form a complete picture of the probability distribution of the particle in space, we must take into consideration the angular dependence of the probability density, which is given by $|Y_{l,m}|^2 = N^2[P_l^{|m|}(\cos\theta)]^2$. The significance of this for the probability density in the two lowest levels is easily seen from the expressions for the spherical harmonics in 4.1. In the $1s$ and $2s$ states the probability distribution is isotropic, and in the three-fold degenerated $2p$-state the angular distribution is proportional to $\cos^2\theta$ or $\sin^2\theta$ for $m = 0$ or $m = \pm 1$, respectively. In the former case the probability density is largest on the z-axis, while in the latter it is largest in the xy-plane.

Here we have only discussed the simplest wave functions; for a more general treatment it is convenient to introduce the above-mentioned generalized Laguerre polynomials. These may be defined by

$$L_p^{(q)}(x) = \frac{\mathrm{d}^q}{\mathrm{d}x^q} L_p(x), \tag{4.3.18}$$

where $L_p(x)$ is the simple Laguerre polynomial

$$L_p(x) = e^x \frac{\mathrm{d}^p}{\mathrm{d}x^p}(x^p e^{-x}), \tag{4.3.19}$$

which apparently is of order p when $L_p^{(q)}$ is of order $p-q$.

The polynomial $L_p^{(q)}$ satisfies Laguerre's differential equation

$$x(L_p^{(q)})'' + (q+1-x)(L_p^{(q)})' + (p-q)L_p^{(q)} = 0, \tag{4.3.20}$$

which is seen to be identical to (4.3.9) with $\lambda = n$ when we set $q = 2l+1$ and $p = n+l$. The polynomial solutions used above can then be written as

$$L(\xi) = L_{n+l}^{(2l+1)}(\xi). \tag{4.3.21}$$

The stationary wave functions for the Coulomb field are thus

$$u_{n,l,m}(r,\theta,\phi) = N_{nl}\xi^l \exp(-\xi/2)\, L_{n+l}^{(2l+1)}(\xi)\, Y_{l,m}(\theta,\phi), \tag{4.3.22}$$

where ξ is given by (4.3.14) and the normalization constant can be shown to be

$$N_{n,l} = \left\{ \left(\frac{2Z}{na_0} \right)^3 \frac{(n-l-1)!}{2n[(n+l)!]^3} \right\}^{\frac{1}{2}}. \tag{4.3.23}$$

Explicit expressions for several of these wave functions are found in Pauling and Wilson, *Introduction to Quantum Mechanics*, McGraw–Hill, New York (1960).

Problems

4.3.1 Find, for a particle in a Coulomb field with main quantum number n and largest possible angular momentum, the most probable distance from the centre of force. Compare this result with the radius of the nth circular orbit according to the 'old quantum mechanics'.

4.3.2 Calculate, for a particle in the ground state in a Coulomb field, the mean value of the potential and the kinetic energy and show that the quantity $Ze^2/\hbar$ is a reasonable estimate for the 'velocity' of the particle.

4.3.3 A negative π-meson ($m_0 = 273\ m_e$) can be captured by a nucleus and form a so-called mesic atom. Evaluate the probability in the states $1s$, $2s$ and $2p$ for the meson to be inside the nucleus (charge of nucleus Z, mass number A, nuclear radius $\approx 1{\cdot}2\ A^{\frac{1}{3}} \times 10^{-13}$ cm) and compare with the corresponding probabilities for an electron.

4.3.4 Find, for a hydrogen atom in the ground state, the distribution of momentum of the electron.

4.3.5 Find, for a particle in the Coulomb potential $-Ze^2/r$, the minimal value of Φ_{eff} and the corresponding r_{min}. Express these quantities by Z, e, a_0 and the lowest value of n corresponding to the l in question.

4.3.6 The following integration formula is valid for Laguerre polynomials:

$$\int_0^\infty \xi^s \exp(-\xi) L_p^{(q)}(\xi) L_{p'}^{(q')}(\xi) d\xi$$

$$= (-1)^{p+p'} p!p'!(s-q)! \sum_{n=0}^{\sigma} \frac{(s+n)!}{n!(p-q-n)!(p'-q'-n)!(s-p+n)!(s-p'+n)},$$

where σ is the smaller of the numbers $p-q$ and $p'-q'$.

Utilize this formula to calculate, for a particle in a Coulomb field, the mean value of $1/r$ for a stationary state with main quantum number n. Show that $\langle 1/r \rangle = Ze^2 m_0/\hbar^2 n^2 = Z/a_0 n^2$, and derive (4.6.1) from this formula.

4.4 ISOTROPIC HARMONIC OSCILLATOR

We now consider a particle of mass m_0 in the spherically symmetric potential

$$\Phi = \tfrac{1}{2}kr^2.$$

Introducing the usual substitution, $R(r) = \chi(r)/r$, the radial equation can be written in the form

$$-\frac{\hbar^2}{2m_0}\chi'' + \left(\tfrac{1}{2}kr^2 + \frac{\hbar^2 l(l+1)}{2m_0 r^2}\right)\chi = E\chi, \tag{4.4.1}$$

and the energy spectrum can be determined by solving (4.4.1) in a similar way as in the previous section. We shall, however, not repeat this method here. Instead it is possible, by using Cartesian coordinates, to reduce the discussion of the three-dimensional oscillator to the one-dimensional case treated in 2.4 [cf. problems 2.4.2 and 4.2.5]. This may be done because the potential $\Phi = \tfrac{1}{2}k(x^2+y^2+z^2)$ can be written as a sum of three one-dimensional oscillator potentials. The stationary Schrödinger equation is then obviously satisfied by

$$u(x, y, z) = u_x(x)\, u_y(y)\, u_z(z), \tag{4.4.2}$$

where the three functions u_x, u_y, u_z are the stationary wave functions for the one-dimensional harmonic oscillator defined in 2.4. The energy E is the sum of the three one-dimensional oscillator energies, which may take the values

$$\omega = \sqrt{\frac{k}{m}}$$

$$\left.\begin{aligned} E_{n_x} &= (\tfrac{1}{2}+n_x)\,\hbar\omega, \qquad & n_x \\ E_{n_y} &= (\tfrac{1}{2}+n_y)\,\hbar\omega, \qquad & n_y \\ E_{n_z} &= (\tfrac{1}{2}+n_z)\,\hbar\omega, \qquad & n_z \end{aligned}\right\} = 0, 1, 2, \cdots. \tag{4.4.3}$$

The energy spectrum for the three-dimensional oscillator is therefore given by

$$E_n = (\tfrac{3}{2}+n)\,\hbar\omega, \quad n = n_x+n_y+n_z = 0, 1, 2, \cdots, \tag{4.4.4}$$

where the quantity n may be called the main quantum number.

By solving the radial equation (4.4.1) we would also have found the energy spectrum (4.4.4) but, would have characterized the eigenstates by the energy E and the quantum numbers l and m, instead of by n_x, n_y and n_z. In the following we shall see how it is possible to find the quantum numbers l and m for a given state without solving (4.4.1), merely on the basis of known properties of the Cartesian solutions (4.4.2). A first hint can be obtained by considering the effective potential in (4.4.1):

$$\Phi_{\text{eff}} = \tfrac{1}{2}kr^2 + \frac{\hbar^2 l(l+1)}{2m_0 r^2}.$$

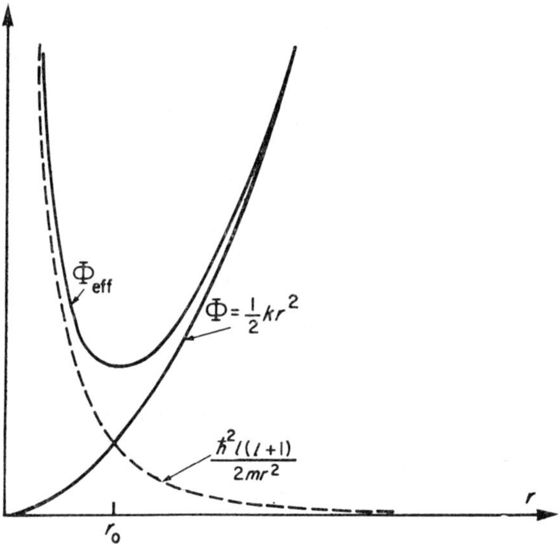

Fig. 4.6

A simple calculation shows that the minimum value of Φ_{eff} (see Fig. 4.6) is

$$(\Phi_{\text{eff}})_{\text{min}} = \Phi_{\text{eff}}(r_0) = \left(\frac{k}{m_0}\right)^{\frac{1}{2}} \hbar\, [l(l+1)]^{\frac{1}{2}} = \hbar\omega[l(l+1)]^{\frac{1}{2}} \quad (4.4.5)$$

and occurs at

$$r = r_0 = \left(\frac{\hbar^2 l(l+1)}{km_0}\right)^{\frac{1}{4}}.$$

For a given value of l the lowest energy level must obviously be above $(\Phi_{\text{eff}})_{\text{min}}$, i.e.

$$E_{n,l} > \hbar\omega[l(l+1)]^{\frac{1}{2}} \geqslant \hbar\omega l.$$

In an energy level with main quantum number n there may thus, according to (4.4.4), only appear states for which

$$l < n + \tfrac{3}{2}. \tag{4.4.6}$$

To obtain additional information about the possible l-values, we may compare the parity of the solutions (4.4.2) with the parity of wave functions of the type $R(r)\,Y_{l,m}(\theta, \phi)$. While the parity of the latter, according to (4.2.18), is even if l is even and odd if l is odd, the parity of the Cartesian solutions is the product of the parity of three functions u_x, u_y and u_z. By the transformation $\mathbf{r} \to -\mathbf{r}$, the wave function (4.4.2) will be multiplied by the factor

$$(-1)^{n_x}\,(-1)^{n_y}\,(-1)^{n_z} = (-1)^n,$$

according to 2.4. Thus all wave functions belonging to the same level have the same parity, which is identical to the parity of n.

The wave functions $R(r)\,Y_{l,m}(\theta, \phi)$ belonging to a definite level can be written as a superposition of solutions (4.4.2) for the same level, from which it follows that there may only appear states where l has the same parity as n. When this is combined with the condition (4.4.6), it is seen that in a level with main quantum number n there may only appear states with

$$l = n,\, n-2, \cdot\cdot\cdot\cdot\cdot,\quad \tfrac{1}{2}[1-(-1)^n]. \tag{4.4.7}$$

Finally, we may determine how many of these allowed l-values actually appear by considering the degree of degeneracy of the levels. This degeneracy can be calculated from the known Cartesian solutions as the number of different sets of quantum numbers n_x, n_y, n_z having the fixed sum n. For n and n_x, fixed, n_y can take on the values from 0 to $n-n_x$, and n_z is then determined. There are thus $n-n_x+1$ solutions for each value of n_x, and the degree of degeneracy is therefore given by

$$\sum_{n_x=0}^{n} (n+1-n_x) = \tfrac{1}{2}\,(n+1)\,(n+2). \tag{4.4.8}$$

For comparison, we calculate the degeneracy on the basis of the angular momentum states under the assumption that all possible values (4.4.7) appear. Each l-state has the degeneracy $2l+1$, and the total degeneracy is therefore given as the sum of a difference series where the largest term is $2n+1$, while the smallest term is 1 or 3 according to whether n is even or odd. Since the number of terms is $(n+2)/2$ for n even and $(n+1)/2$ for n odd, one finds again for all n the degeneracy (4.4.8).

Thus to get the correct degree of degeneracy we must admit all the possible angular momentum states (4.4.7), and may associate the azimuthal quantum numbers with the energy levels as shown in Fig. 4.7.

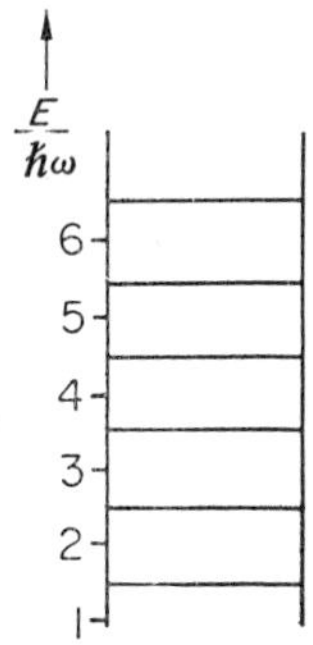

$n = n_x + n_y + n_z$	l	Degeneracy	Notation
5	1, 3, 5	21	4p 5f 6h
4	0, 2, 4	15	3s 4d 5g
3	1, 3	10	3p 4f
2	0, 2	6	2s 3d
1	1	3	2p
0	0	1	1s

Fig. 4.7

It should be noted that the main quantum number n bears no direct relation to the main quantum number discussed in the preceding section. The radial quantum number is indicated in the last column of Fig. 4.7 and is determined by the sequence of the energy levels for each l-value.

Problems

4.4.1 Determine from (4.4.1) the radial functions $R(r)$ belonging to all s-states, and find the corresponding energy values.

4.4.2 Determine the energy spectrum for the isotropic harmonic oscillator by solving the radial equation (4.4.1). Find first asymptotic solutions and then introduce $\chi(r)$ as the product of these and a power series.

4.5 THE FREE PARTICLE AND THE SPHERICAL POTENTIAL WELL

In the description of the motion of a nucleon in an atomic nucleus under the influence of nuclear forces, one often uses as a first approximation a spherical potential well. The width of the well represents the size of the nucleus, and the depth corresponds to the work performed by the nuclear forces when the nucleon moves through the nuclear surface. In the following we shall, for simplicity, limit the discussion to a potential well with infinitely high walls.

We assume $\Phi = 0$ for $r < a$, and after the usual substitution $R = \chi/r$, we may write the radial equation in the form

$$-\frac{\hbar^2}{2m_0}\chi'' + \frac{\hbar^2 l(l+1)}{2m_0 r^2}\chi = E\chi. \tag{4.5.1}$$

In order to determine the stationary states we must find solutions to (4.5.1) which, besides vanishing at the origin (so that R remains finite), also satisfy $\chi(a) = 0$ for $r = a$, where the potential is assumed to jump to infinity.

If one uses the substitutions

$$\xi = kr, \qquad k = \frac{(2m_0 E)^{\frac{1}{2}}}{\hbar} \tag{4.5.2}$$

and

$$\chi(r) = \xi^{\frac{1}{2}} J(\xi), \tag{4.5.3}$$

the radial equation (4.5.1) may be written

$$\xi^2 \frac{d^2 J}{d\xi^2} + \xi \frac{dJ}{d\xi} + [\xi^2 - (l+\tfrac{1}{2})^2]\, J = 0. \tag{4.5.4}$$

This is known as the Bessel differential equation; it has as independent solutions the Bessel functions of order $p = l+\frac{1}{2}$ and $p = -(l+\frac{1}{2})$, but since Bessel functions $J_p(\xi)$, of order p, behave like ξ^p for $\xi \to 0$*, it is seen that the acceptable solutions must be of the form

$$J = J_{l+\frac{1}{2}}(\xi) = J_{l+\frac{1}{2}}(kr). \tag{4.5.5}$$

This is a solution of (4.5.1), but in order to satisfy the boundary conditions at $r = a$, k must satisfy the equation $J_{l+\frac{1}{2}}(k\,a) = 0$. For each zero of the Bessel function $J_{l+\frac{1}{2}}(\xi)$ one thus finds a possible value of k, and from (4.5.2) one may then determine an energy level belonging to the azimuthal quantum number l in question. The number of the zero determines the radial quantum number.

In Fig. 4.8. the first three zeros of $J_p(\xi)$ are illustrated as a function

* This holds if p is not a negative integer.

of p^*. The crosses mark points which lead to possible stationary states. To the right the corresponding ten lowest energy levels are indicated as they are obtained from the zeros ξ_0 by the relation

$$E = \frac{\hbar^2 \xi_0{}^2}{2m_0 a^2}. \tag{4.5.6}$$

It is noted that the energy levels in this case do not show any degeneracy in l, in contrast to what we found for the Coulomb field and for the harmonic oscillator. The energy levels have here the minimal degeneracy for a central field, namely $2l+1$.

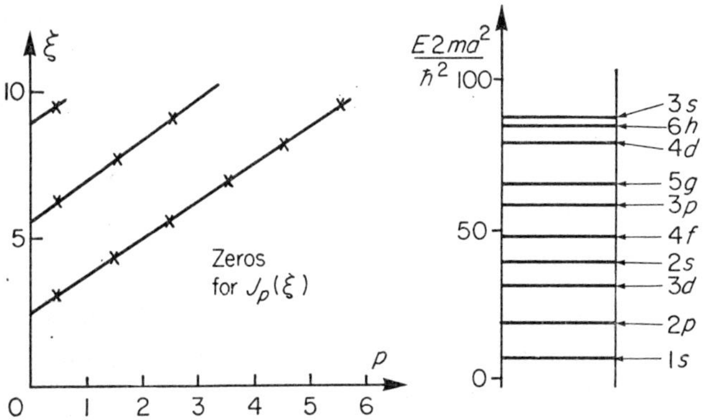

Fig. 4.8

As was seen above, the radial functions can be written in the form

$$R_l = N_l j_l(kr), \tag{4.5.7}$$

where

$$j_l(\xi) = \left(\frac{\pi}{2\xi}\right)^{\frac{1}{2}} J_{l+\frac{1}{2}}(\xi) \tag{4.5.8}$$

are the so-called spherical Bessel functions. These Bessel functions have the special feature that they can be expressed by means of elementary functions. One may thus show [cf. problem 4.5.1] that the spherical Bessel functions can be written

$$j_l(\xi) = \xi^l \left(-\frac{1}{\xi}\frac{\mathrm{d}}{\mathrm{d}\xi}\right)^l \frac{\sin \xi}{\xi}. \tag{4.5.9}$$

* Accurate values for the zeros can be found, for instance in *Handbook of Mathematical Functions,* National Bureau of Standards, Applied Mathematics Series no. 55, Washington, D.C. (1964).

M

The radial functions for *s*-, *p*- and d-states are thus found to be

$$R_0 = N_0 \frac{\sin kr}{kr},$$

$$R_1 = N_1 \left(\frac{\sin kr}{(kr)^2} - \frac{\cos kr}{kr}\right), \tag{4.5.10}$$

$$R_2 = N_2 \left[\left(\frac{3}{(kr)^3} - \frac{1}{kr}\right)\sin kr - 3\frac{\cos kr}{(kr)^2}\right].$$

The factors N_l are determined from the normalization condition

$$\int_0^a [R_l(kr)]^2 r^2 \mathrm{d}r = 1. \tag{4.5.11}$$

In these radial functions one must insert one of the k-values which lead to $R(k\,a) = 0$, and each function thus represents infinitely many wave functions corresponding to radial quantum numbers $n_r = 1, 2, \cdots$.

The radial equation (4.5.1) is identical to the radial equation for a free particle. For a free particle we must also satisfy the regularity condition at $r = 0$, and we thus find the same stationary wave functions as above:

$$u(\mathbf{r}) = N_l j_l(kr) Y_{lm}(\theta, \phi). \tag{4.5.12}$$

We may, however, in this case choose k freely. The wave functions (4.5.12) are improper. An appropriate normalization can be determined, e.g. from the condition (4.5.11) through a limiting process $a \to \infty$ [see equation (3.7.10)].

While the wave function for the free particle in Cartesian coordinates is characterized by the values of the three components of the linear momentum, the wave function (4.5.12) represents a state where the angular momentum with respect to the origin has a definite value. Since the states (4.5.12) form a complete orthogonal system it is, of course, possible to expand for instance a plane wave with definite linear momentum over these wave functions. For a plane wave with momentum $p = k\hbar$ along the z-axis one finds the result

$$\exp(\mathrm{i}kz) = \sum_{l=0}^{\infty} \mathrm{i}^l (2l+1) j_l(kr) P_l(\cos\theta). \tag{4.5.13}$$

From this formula it is seen how the plane wave contains all possible angular momenta around the origin. In accordance with the physical significance, all angular momenta are perpendicular to the z-axis ($m = 0$).

Problems

4.5.1 Write $\chi(r) = r^{l+1}f(r)$ and find, from (4.5.1), the differential equation for f. Show then by induction that (4.5.1) is satisfied by

$$\chi(r) = r^{l+1}\left(\frac{1}{r}\frac{\mathrm{d}}{\mathrm{d}r}\right)^l \frac{\sin kr}{r}.$$

Compare this result with (4.5.9).

4.5.2 Determine the bound s-states in a spherical potential well with radius a and finite depth Φ_0. Compare the result with the levels of the one-dimensional well. Show that, for $\Phi_0 a^2 < \pi^2\hbar^2/8m_0$, there are no bound s-states in the spherical well. If $\Phi_0 a^2$ is slightly larger than the above limit there is just one level; find for this case an approximate expression for the binding energy.

4.5.3 In problem 4.5.2 it was shown that there are no bound s-states in a potential well where the depth Φ_0 and the radius a satisfy the inequality $\Phi_0 a^2 < \pi^2\hbar^2/8m_0$. Show that there are in this case also no bound states with $l > 0$. A particle can thus not be bound at all in a sufficiently small spherical well. Why is this not the case for a one-dimensional well?

4.5.4 Find, from (4.5.13), the relative probability of having the angular momentum l in a plane wave $\exp(ikz)$.

4.6 THE HYDROGEN ATOM—DEVIATIONS FROM THE SIMPLE DESCRIPTION

As a conclusion of the discussion of the bound states of a particle in a central field, we will discuss some effects which were neglected above but which turn out to be of interest in actual atomic systems. In particular, we will discuss the applicability of the result of 4.3 to a hydrogen-like atom.

The simplest effect is connected with the fact that for hydrogen-like atoms we are concerned with a two-body problem, and not with the motion of a single particle in a static field. A discussion of the reduction of the two-body problem is given in 5.1. We may here note, as was already mentioned in 4.1, that the result of this reduction is that one should replace the electron mass by the slightly smaller reduced mass. It is well known that the introduction of the reduced mass leads to small differences between the Rydberg constants for the spectra of H, He^{+}, Li^{2+}, etc. Although the effect is small because of the large mass ratio between the nucleus and the electron, it is easily observed in spectrographic measurements. A much larger effect is found for mesic atoms where a negative π- or μ-meson is bound in the Coulomb field of a nucleus [cf. problem 5.1.1.].

One further important difference between normal and mesic atoms arises from the finite size of the nucleus. The stationary states studied in 4.3 pertain to a pure Coulomb field where the nucleus is assumed to be a point charge. It is more appropriate to consider the nucleus as a smeared charge distribution, which leads to considerable deviations from the Coulomb field in a region of the order of magnitude of the size of the nucleus.* This implies that the energy levels will depend on the charge distribution in the nucleus, so that, for example, the energy levels in hydrogen and deuterium are slightly different, and this correction is to be added to the effect of the reduced mass. This so-called isotope effect is quite small for normal atoms since the radius of the nucleus is 10^2–10^4 times smaller than the extent of the electronic wave function, and the effect is noticeable only for s-states, where the probability density is finite at $r = 0$. For mesic atoms the effect is quite large. In fact, meson masses are 200–300 times larger than the electron mass; the quantity $a_0/Z = \hbar^2/Zm_0e^2$, which is the measure of the extent of the wave function in the ground state, may then be of the same order of magnitude as the nuclear radius. Only in the case of high quantum numbers and low nuclear charge will the pure Coulomb levels give a good description of a mesic atom [cf. equation (4.3.17)]. In return, the experimentally observed lower levels yield information about the charge distribution in the nucleus.

We considered above a few effects for which the treatment was quite straightforward. We shall now discuss some basic deficiencies of our description. These deficiencies are primarily that the treatment was unrelativistic and that we have disregarded spin.

In order to estimate the error of a non-relativistic calculation we shall, for a hydrogen-like atom, compare the rest energy mc^2 with the expectation value, $\langle T\rangle$, of the non-relativistic kinetic energy.

---

* For $r\to 0$ one thus finds that $\Phi(\mathrm{r})$ remains finite for the smeared charge distribution.

The ratio between these two quantities is found to be [cf. problem 4.3.6]

$$\frac{\langle T \rangle}{mc^2} = \frac{Z^2 e^4}{2\hbar^2 c^2 n^2} = \frac{1}{2}\frac{Z^2}{n^2} \cdot \alpha^2, \tag{4.6.1}$$

where

$$\alpha = \frac{e^2}{\hbar c} = \frac{1}{137} \tag{4.6.2}$$

is the so-called fine structure constant introduced by Sommerfeld. According to (4.6.1), the relative error in the energy must be of the order of $Z^2\alpha^2/n^2$ which, for a hydrogen atom in the ground state, implies an error of about 10^{-4}. For higher levels the error diminishes rapidly, while it increases as Z^2, so that one makes a considerable error by applying a non-relativistic treatment in the case of the innermost orbits in a heavy atom.

We now turn to the effect of spin. If the particle in the central field has an intrinsic angular momentum (cf. 5.3), we cannot, of course, immediately consider the orbital angular momentum as a constant of the motion: this rôle is taken over by the total angular momentum. Spin and orbital angular momentum could only be separate constants of the motion if there were no coupling between them. In an atom, however, there is a coupling of this kind, since the orbital motion and spin of an electron create magnetic moments which have a mutual interaction. The interaction energy of the spin–orbit coupling in an atom is of the same order of magnitude as the relativistic corrections to the energy levels. In an accurate theory of the electronic structure of atoms one must therefore introduce both relativistic corrections and spin corrections. A proper way of treating these corrections is to use Dirac's celebrated relativistic wave equation for the electron. In this equation the spin effects appear directly as a consequence of relativistic invariance, and therefore need not be introduced as extra corrections. As an example of the results of Dirac's theory, we quote the relativistic expression for the energy levels of a particle (with spin $\frac{1}{2}$) in a Coulomb field:

$$E_{n,j} = mc^2 \left[1 + \frac{Z^2\alpha^2}{[n-(j+\frac{1}{2}) + \sqrt{(j+\frac{1}{2})^2 - Z^2\alpha^2}]^2} \right]^{-\frac{1}{2}}, \tag{4.6.3}$$

where α is the fine structure constant (4.6.2), while the main quantum number n can take on the values 1, 2, 3 $\cdots$. The quantum number $j = \frac{1}{2}, \frac{3}{2}, \cdots, n-\frac{1}{2}$ determines the magnitude of total angular momentum, $\sqrt{j(j+1)}\hbar$.

We have here introduced spin in association with relativistic corrections. It should, however, be noted that the two effects may usually be considered separately. On the one hand, the relativistic correction in itself is of interest for a π-mesic atom because the π-meson has no spin. On the other hand, the spin correction is of significance in the description of nuclear structure because the non-central nature of nuclear forces introduces strong additional spin–orbit couplings. For the electrons, too, one often uses non-relativistic theory with spin because in this way one can account for many features of the electronic structure of atoms (e.g. the Pauli exclusion principle, 5.5).

Finally we shall touch upon a grave deficiency in our description in as far as it is applied to electrically charged particles. This deficiency is our neglect of electromagnetic radiation from the system. The energy spectrum for a hydrogen atom, as treated in 4.3, gives with good accuracy the values of the terms from which one may construct the emission spectrum by means of Bohr's frequency condition. The energy values belong to the states which were called stationary in the old quantum mechanics. In the above we have also treated them as stationary, but strictly speaking this can only hold for the ground state, because empirically an atom can only remain in an excited state for a limited time whereupon it emits radiation. In a more detailed description the excited states must be considered as quasi-stationary and, in fact, one also finds finite widths for the emitted spectral lines which reflect the finite widths of the quasi-stationary levels.

We were led to consider the excited states of the hydrogen atom as stationary because we neglected the interaction between the charged particle and the electromagnetic field. It is obvious that if the atom was subjected to an external radiation field this interaction should be taken into consideration, and an absorption can then also be described by the above theory if one introduces a term in the Hamilton operator accounting for the action of the electromagnetic wave on the electron. So far we have, however, considered an isolated atom without external fields, and it is then less obvious that in 'empty space' one should include an interaction term which can give rise to a so-called spontaneous emission. Nonetheless, this is reasonable, as may be seen in various ways.

We know, for example, from classical physics, that a system of charged particles, even though it is completely isolated from the surroundings, cannot, in general, conserve its mechanical energy. The latter will be reduced by the emission of electromagnetic radiation, and this is exactly what is observed for the excited states of an atom. For our quantum mechanical description, the possibility of emission of radiation means, however, that there is a—so far ignored—'channel' open for the decay of the excited state. As a result the state can, at most, be quasi-stationary. True stationary excited states could only be expected if one could enclose the atom and the radiation in a box so that a stationary state would consist of a superposition of atomic states and radiation.

In a more detailed treatment of radiation processes the electromagnetic field must also be treated quantum mechanically (as a photon system). In the initial state the atom is supposed to be in a definite excited state (mechanically stationary) whereas the field consists of zero photons. The interaction, which can be treated as a small perturbation, brings about in course of time a change of the mechanical state of the atom accompanied by creation of photons, so that the total energy is conserved. The new state of the atom may be regarded as a superposition of the previously found 'stationary' states, and thus one finds the picture (cf. 3.8) of a decay process as a transition between stationary states. Initially the atom was in one state, but at a later time it is found with a certain probability in various other 'stationary' states. Energy conservation is taken into account in the way that the atom can only make transitions to lower states accompanied by emission of photons with corresponding probabilities and energies.

Problems

4.6.1 Perform a series expansion of (4.6.3) in powers of $Z\alpha$ up to terms of order of $Z^4\alpha^4$. Show, for $Z\alpha \to 0$, that there is agreement with the non-relativistic expression for the energy levels in a Coulomb field. Show that in next approximation the energy terms (4.3.13) are corrected by a factor

$$1+\frac{Z^2\alpha^2}{n^2}\left(\frac{n}{j+\frac{1}{2}}-\frac{3}{4}\right).$$

Compute from this the fine structure splitting of the $n = 2$ and $n = 3$ levels in the hydrogen atom.

4.6.2 Estimate, by means of lowest order perturbation theory, the isotope shift between the ground states of a hydrogen and a deuterium atom. (The hydrogen nucleus is considered as a point charge while the charge distribution in deuterium is assumed to be constant within a radius of 5×10^{-13} cm.)

5

Systems Containing Several Particles

In quantum mechanics, as in classical mechanics, the description of the motion of a single particle provides the natural basis for the description of particle systems. For a system of several particles one proves in classical mechanics a number of theorems which give a better understanding of the notion 'a particle'. For a non-relativistic motion one may thus show that the kinetic energy for a system of particles can be written as the sum of the particle energies in the centre of mass system plus the the kinetic energy of a particle with the mass of the total system moving with the velocity of the centre of mass (theorem of König).

For an isolated system this theorem allows a complete separation of the intrinsic motion of the system from the motion of the system as a whole. In the intrinsic motion the energy, as well as the total angular momentum with respect to the centre of the mass, are conserved quantities.

In a large number of cases it is possible to maintain an approximate separation also for systems which interact with the surroundings. In such cases the motion of the centre of mass is the same as it would be if the system were a particle, and the structure of the system is only revealed through the mass. It is, in this connexion, important to note that this also applies if the intrinsic motion is relativistic. The total energy of an isolated system of particles is in general given by the expression

$$E = \frac{Mc^2}{(1-v^2/c^2)^{\frac{1}{2}}},$$

where v is the velocity of the centre of mass. Furthermore,

$$Mc^2 = \sum m_i c^2 + E_{\text{int}}$$

where E_{int} is the sum of the kinetic and potential energies in the centre of mass system. In the limit $v \ll c$, the above equation may be written

$$E - \sum m_i c^2 = \tfrac{1}{2}\left(\sum m_i + \frac{E_{\text{int}}}{c^2}\right)v^2 + E_{\text{int}},$$

which is indentical to the theorem of König, if $E_{\text{int}} \ll Mc^2$. We thus see that even if this condition is not fulfilled we may, in the non-relativistic motion of our system as a whole, include a possibly relativistic intrinsic motion by adding the corresponding energy to the mass of the 'particle'.

In many cases, external forces on the system will be such that the direction of the intrinsic angular momentum of the system can be changed by interaction with the surroundings under conservation of the intrinsic energy (compare, for example, the motion of a spinning top). In such cases it is natural to consider the system as a particle, which, besides possessing a definite mass also has an intrinsic angular momentum, the direction of which may be changed by external forces.

In this chapter we shall first discuss how such phenomena are to be formulated in quantum mechanics. Towards the end of the chapter we shall discuss the intrinsic motion in many-body problems.

5.1 THE SCHRÖDINGER EQUATION FOR SEVERAL PARTICLES

The formal description of several particles being a straightforward generalization of the theory for a single particle, we shall here only summarize the quantum theory of particle systems.

We shall introduce a wave function, ψ, of a particle system, which depends on all the particle coordinates such that the absolute square $|\psi|^2$ indicates the probability density in configuration space. To a system of N particles with coordinates $\mathbf{r}_1, \mathbf{r}_2, \mathbf{r}_3, \ldots \mathbf{r}_N$, one associates a wave function $\psi(\mathbf{r}_1, \mathbf{r}_2, \mathbf{r}_3, \ldots, \mathbf{r}_N, t)$ and the quantity

$$P(\mathbf{r}_1, \mathbf{r}_2, \ldots \mathbf{r}_N, t)\, d\tau_1\, d\tau_2 \cdots d\tau_N \tag{5.1.1}$$

$$= |\psi(\mathbf{r}_1, \ldots \mathbf{r}_N, t)|^2\, d\tau_1 \cdots d\tau_N$$

indicates the probability at time t of finding the N particles in the volume elements $d\tau_1, d\tau_2, \ldots . d\tau_N$ at the positions $\mathbf{r}_1, \mathbf{r}_2, \ldots . \mathbf{r}_N$, respectively. From this interpretation follows the normalization condition

$$\int |\psi(\mathbf{r}_1, \ldots . \mathbf{r}_N, t)|^2 \, d\tau_1 \, d\tau_2 \cdots d\tau_N = 1. \tag{5.1.2}$$

The Schrödinger equation for a particle system can formally be written as for a single particle:

$$\hat{H}\psi = i\hbar \frac{\partial \psi}{\partial t}. \tag{5.1.3}$$

The Hamilton operator, $\hat{H}$, is constructed from the classical Hamiltonian for a particle system by substituting the momenta $\mathbf{p}_1, \mathbf{p}_2, \ldots .$ of all the particles by the momentum operators

$$\hat{\mathbf{p}}_1 = -i\hbar\left(\frac{\partial}{\partial x_1}, \frac{\partial}{\partial y_1}, \frac{\partial}{\partial z_1}\right),$$

$$\hat{\mathbf{p}}_2 = -i\hbar\left(\frac{\partial}{\partial x_2}, \frac{\partial}{\partial y_2}, \frac{\partial}{\partial z_2}\right), \text{ etc.}$$

For a system of N particles with potential energy

$$\phi = \phi\,(\mathbf{r}_1, \mathbf{r}_2, \ldots, \mathbf{r}_N) \tag{5.1.4}$$

the Schrödinger equation (5.1.3) thus takes the form

$$\left[\sum_{i=1}^{N}\left(-\frac{\hbar^2}{2m_i}\Delta_i\right) + \Phi(\mathbf{r}_1, \mathbf{r}_2, \ldots, \mathbf{r}_N)\right]\psi(\mathbf{r}_1, \cdots, \mathbf{r}_N, t)$$

$$= i\hbar\frac{\partial}{\partial t}\psi(\mathbf{r}_1, \mathbf{r}_2, \ldots \mathbf{r}_N, t), \tag{5.1.5}$$

where m_i denotes the mass of the ith particle and Δ_i is the Laplacian operator,

$$\Delta_i = \frac{\partial^2}{\partial x_i^{\,2}} + \frac{\partial^2}{\partial y_i^{\,2}} + \frac{\partial^2}{\partial z_i^{\,2}},$$

acting only on the coordinates of the ith particle. If the wave function, $\psi(\mathbf{r}_1, \mathbf{r}_2, \cdots, \mathbf{r}_N, t_0)$, is known at the time t_0, one may determine it at any later time by means of equation (5.1.5).

Expectation values of physical quantities are given in a similar way as for a single particle. Thus the expectation value of the energy is given by

$$\langle E\rangle = \int \psi^*(\mathbf{r}_1 \ldots, \mathbf{r}_N, t)\, \hat{H}\psi(\mathbf{r}_1, \ldots, \mathbf{r}_N, t)\, d\tau_1 \cdots d\tau_N, \tag{5.1.6}$$

where the integration is extended over the entire $3N$-dimensional configuration space.

Special solutions of the Schrödinger equation (5.1.3) are obtained from the separation ansatz (compare 2.1):

$$\psi(\mathbf{r}_1, \ldots \mathbf{r}_N, t) = u(\mathbf{r}_1, \ldots \mathbf{r}_N) \exp(-iEt/\hbar), \tag{5.1.7}$$

where u is independent of t. One finds that (5.1.7) satisfies (5.1.3) if u is the solution of the time-independent Schrödinger equation

$$\hat{H}u = Eu. \tag{5.1.8}$$

Evidently the solutions (5.1.7) describe stationary states where the total energy has the well-defined value E. Solutions to (5.1.8) belonging to two different energy values E_1 and E_2 satisfy the orthogonality relations (compare 2.5):

$$\int u^*_{E_1}(\mathbf{r}_1, \ldots \mathbf{r}_N)\, u_{E_2}(\mathbf{r}_1, \ldots \mathbf{r}_N)\, d\tau_1 \cdots d\tau_N = 0. \tag{5.1.9}$$

In the continuous spectrum the wave functions are improper. As before, we shall assume that the set of all solutions to (5.1.8) form a complete orthogonal system in the variables $\mathbf{r}_1, \mathbf{r}_2, \ldots, \mathbf{r}_N$.

As a simple example of a system of particles we consider N particles moving independently of each other. For such a system the potential energy is given by

$$\Phi = \Phi_1(\mathbf{r}_1) + \Phi_2(\mathbf{r}_2) + \ldots + \Phi_N(\mathbf{r}_N), \tag{5.1.10}$$

where Φ_i depends only on $\mathbf{r}_i$. The total Hamilton operator can thus be written

$$\hat{H} = \sum_i \hat{H}_i, \tag{5.1.11}$$

where

$$\hat{H}_i = -\frac{\hbar^2}{2m_i}\Delta_i + \Phi_i(\mathbf{r}_i). \tag{5.1.12}$$

In this case the stationary Schrödinger equation (5.1.8) can be solved by separation. Assuming that

$$u = u_1(\mathbf{r}_1)u_2(\mathbf{r}_2) \cdots u_N(\mathbf{r}_N), \tag{5.1.13}$$

one finds that (5.1.13) is a solution of (5.1.8) if u_i satisfies the Schrödinger equation for a single particle in the potential Φ_i:

$$\hat{H}_i u_i(\mathbf{r}_i) = E_i u_i(\mathbf{r}_i), \tag{5.1.14}$$

and one finds furthermore that the energy eigenvalue E is given by the sum of the single particle energies

$$\sum_i E_i = E. \tag{5.1.15}$$

These results are in complete accordance with our expectations. In fact the results (5.1.13)–(5.1.15) should rather be taken as a condition to be fulfilled by the formalism of quantum mechanics for several particles which was given above. The one-particle description, if valid, demands that when a number of particles have no mutual interaction, eigenstates of the energy of each particle lead to an energy eigenstate of the total system with the eigenvalue (5.1.15). It should also be possible to write the probability distribution

$$\begin{aligned} &P(\mathbf{r}_1, \mathbf{r}_2, \ldots, \mathbf{r}_N) \\ &= |u_1(\mathbf{r}_1)|^2 |u_2(\mathbf{r}_2)|^2 \cdot \cdots \cdot |u_N(\mathbf{r}_N)|^2 \\ &= P_1(\mathbf{r}_1)\, P_2(\mathbf{r}_2) \cdot \cdots \cdot P_N(\mathbf{r}_N) \end{aligned} \tag{5.1.16}$$

as a product of the probabilities $P_i(\mathbf{r}_i)$ of finding the particle i at the position $\mathbf{r}_i$, corresponding to the fact that the particles move independently.

As another important example, we consider that of an isolated system of N particles where we expect, as mentioned above, that one may separate the centre of mass motion from the intrinsic motion. Let us commence the discussion by considering a two-particle problem. The Hamilton operator is here assumed to be of the form

$$\hat{H} = -\frac{\hbar^2}{2m_1}\Delta_1 - \frac{\hbar^2}{2m_2}\Delta_2 + \Phi(\mathbf{r}_1 - \mathbf{r}_2), \tag{5.1.17}$$

where Φ only depends on the relative position vector.

If we perform a coordinate transformation where we introduce, instead of the position vectors $\mathbf{r}_1$ and $\mathbf{r}_2$, the relative position

$$\mathbf{r}_{12} = \mathbf{r}_1 - \mathbf{r}_2, \tag{5.1.18}$$

and the centre of mass coordinates

$$\mathbf{R}_{12} = \frac{m_1\mathbf{r}_1 + m_2\mathbf{r}_2}{m_1+m_2}, \tag{5.1.19}$$

one finds that

$$\begin{aligned} \Delta_1 &= \Delta_r + \left(\frac{m_1}{M}\right)^2 \Delta_R + \frac{2m_1}{M} \nabla_r \cdot \nabla_R \\ \Delta_2 &= \Delta_r + \left(\frac{m_2}{M}\right)^2 \Delta_R - \frac{2m_2}{M} \nabla_r \cdot \nabla_R, \end{aligned} \tag{5.1.20}$$

where $\Delta_r = (\nabla_r)^2$ and $\Delta_R = (\nabla_R)^2$ are the Laplace operators in the coordinates $\mathbf{r}_{12}$ and $\mathbf{R}_{12}$ respectively, while

$$M = m_1 + m_2 \tag{5.1.21}$$

is the total mass.

Inserting (5.1.20) into (5.1.17) we find that

$$\hat{H} = -\frac{\hbar^2}{2m_0} \Delta_r - \frac{\hbar^2}{2M} \Delta_R + \Phi(\mathbf{r}_{12}), \tag{5.1.22}$$

where

$$m_0 = \frac{m_1 m_2}{m_1+m_2} \tag{5.1.23}$$

is the reduced mass.

We now consider the wave function u as a function of $\mathbf{r}_{12}$ and $\mathbf{R}_{12}$. The Schrödinger equation

$$\hat{H}u(\mathbf{r}_{12}, \mathbf{R}_{12}) = Eu(\mathbf{r}_{12}, \mathbf{R}_{12}) \tag{5.1.24}$$

is then of the form (5.1.11)–(5.1.12), which can be solved by the separation ansatz

$$u(\mathbf{r}_{12}, \mathbf{R}_{12}) = u_{\text{int}}(\mathbf{r}_{12}) \cdot u_g(\mathbf{R}_{12}) \tag{5.1.25}$$

where the two functions u_{int} and u_g satisfy the equations

$$\hat{H}_{\text{int}} u_{\text{int}}(\mathbf{r}_{12}) = \left[-\frac{\hbar^2}{2m_0} \Delta_r + \Phi(\mathbf{r}_{12})\right] u_{\text{int}}(\mathbf{r}_{12}) = E_{\text{int}} u_{\text{int}}(\mathbf{r}_{12}) \tag{5.1.26}$$

and

$$\hat{H}_g u_g(\mathbf{R}_{12}) = \left(-\frac{\hbar^2}{2M} \Delta_R\right) u_g(\mathbf{R}_{12}) = E_g u_g(\mathbf{R}_{12}), \tag{5.1.27}$$

while the energy E satisfies

$$E = E_{\text{int}} + E_g. \tag{5.1.28}$$

We have thus proved that the two-particle problem can be separated into an independent, free motion of the centre of mass (5.1.27), and an intrinsic relative motion (5.1.26) of a reduced particle in the field $\Phi(\mathbf{r}_{12})$. All the results of the previous chapters for the motion of a single particle can therefore be interpreted in terms of the relative motion of two particles interacting through the potential $\Phi(\mathbf{r})$, merely by substituting for m the reduced mass of the two particles. The total wave function for the two-particle problem is obtained by multiplying the relative wave function by a wave function describing the free motion of the centre of mass, e.g. a plane wave,

$$u_g(\mathbf{R}_{12}) = \exp(\mathrm{i}\mathbf{k}\mathbf{R}_{12}), \tag{5.1.29}$$

where the magnitude k of the wave vector $\mathbf{k}$ is given by

$$k = \left(\frac{2ME_g}{\hbar^2}\right)^{\frac{1}{2}}. \tag{5.1.30}$$

The wave function (5.1.29) is obviously an eigenstate for the momentum operator

$$\hat{\mathbf{P}} = -\mathrm{i}\hbar\nabla_R \tag{5.1.31}$$

with eigenvalue

$$\mathbf{P} = \hbar\mathbf{k}. \tag{5.1.32}$$

Using the definitions (5.1.18)–(5.1.19), one easily verifies that the operator (5.1.31) may be written

$$\hat{\mathbf{P}} = \hat{\mathbf{p}}_1 + \hat{\mathbf{p}}_2, \tag{5.1.33}$$

where $\hat{\mathbf{p}}_1$ and $\hat{\mathbf{p}}_2$ are the momentum operators of the two particles, and $\hat{\mathbf{P}}$ thus indicates the total momentum of the two-particle system. Since $\hat{\mathbf{P}}$ [see equation (5.1.31)] commutes with $\hat{H}$ [see equation (5.1.22)], it is a constant of the motion.

The operator

$$\hat{\mathbf{p}}_{12} = -\mathrm{i}\hbar\nabla_r \tag{5.1.34}$$

indicates the momentum in the relative motion. As can be verified from the definitions (5.1.18)–(5.1.19), it can be written

$$\hat{\mathbf{p}}_{12} = m_0\left(\frac{\hat{\mathbf{p}}_1}{m_1} - \frac{\hat{\mathbf{p}}_2}{m_2}\right). \tag{5.1.35}$$

The momentum $\hat{\mathbf{p}}$ commutes with $\hat{\mathbf{P}}$, but not usually with $\hat{H}$.

In most cases the potential $\Phi(\mathbf{r}_{12})$ will only depend on the magnitude r_{12} of $\mathbf{r}_{12}$, and the relative motion thus becomes equivalent to the motion of a particle in a spherically symmetric field. In this case the relative angular momentum

$$\hat{\mathbf{S}} = \mathbf{r}_{12} \times \hat{\mathbf{p}}_{12} \tag{5.1.36}$$

is a constant of the motion. Besides commuting with $\hat{H}$ it also commutes with $\hat{\mathbf{P}}$ and we may choose $\hat{H}_{\text{int}}$, $\hat{\mathbf{S}}^2$, $\hat{S}_z$ and $\hat{\mathbf{P}}$ as a (complete) set of commuting operators for the two-particle system. Also the angular momentum associated with the centre of mass motion

$$\hat{\mathbf{L}} = \mathbf{R}_{12} \times \hat{\mathbf{P}} \tag{5.1.37}$$

is a constant of the motion which commutes with $\hat{\mathbf{S}}$. We could therefore instead choose six operators $\hat{H}$, $\hat{H}_{\text{int}}$, $\hat{S}^2$, $\hat{S}_z$, $\hat{L}^2$ and $\hat{L}_z$ as a basis where the eigenstates for the centre of mass motion are of the form discussed in chapter 4.

It follows directly from the equations (5.1.33)–(5.1.37) that the sum of $\hat{\mathbf{S}}$ and $\hat{\mathbf{L}}$ is equal to the total angular momentum of the two particles, i.e.

$$\hat{\mathbf{S}} + \hat{\mathbf{L}} = \hat{\mathbf{J}}, \tag{5.1.38}$$

where

$$\hat{\mathbf{J}} = \mathbf{r}_1 \times \hat{\mathbf{p}}_1 + \mathbf{r}_2 \times \hat{\mathbf{p}}_2. \tag{5.1.39}$$

It is therefore natural to call $\mathbf{S}$ the intrinsic angular momentum.

The total angular momentum is obviously also a constant of the motion. However, it should be noted that since $|\hat{\mathbf{J}}|^2$ does not commute with $\hat{S}_z$ and $\hat{L}_z$, the states mentioned above, where $\hat{S}^2$, $\hat{S}_z$, $\hat{L}^2$ and $\hat{L}_z$ have definite values, are not eigenstates of the total angular momentum. The problem of constructing such states will be discussed in 5.6 below, and we shall now instead return to the problem of separating the centre of mass motion for more than two particles.

This is quite easy once we know the result for the two-particle problem, since we can generate the coordinate transformation in steps, where in each step only two of the particle coordinates are reduced to relative particle coordinates and centre of mass coordinates. Thus if we consider three particles with positions $\mathbf{r}_1$, $\mathbf{r}_2$, $\mathbf{r}_3$ and masses m_1, m_2, m_3 we may first change from the coordinates $\mathbf{r}_1$, $\mathbf{r}_2$ and $\mathbf{r}_3$ to the coordinates $\mathbf{r}_{12}$, $\mathbf{R}_{12}$ and $\mathbf{r}_3$ by means of the transformations (5.1.18)–

(5.1.19). Next we perform the analogous transformation from $\mathbf{R}_{12}$ and $\mathbf{r}_3$ to the relative coordinate $\mathbf{r}_{3(12)}$ of particle 3 with respect to the centre of mass $\mathbf{R}_{12}$ of particles 1 and 2, and the centre of mass $\mathbf{R}_{123}$ of these two systems, which is in fact the centre of mass of all three particles (see Table 5.1).

Table 5.1

riginal			
ass	m_1	m_2	m_3
oordinate	$\mathbf{r}_1$	$\mathbf{r}_2$	$\mathbf{r}_3$
ntermediate			
ass	$m_{12} = \dfrac{m_1 m_2}{m_1+m_2}$	$M_{12} = m_1+m_2$	m_3
oordinate	$\mathbf{r}_{12} = \mathbf{r}_1-\mathbf{r}_2$	$\mathbf{R}_{12} = \dfrac{m_1\mathbf{r}_1+m_2\mathbf{r}_2}{m_1+m_2}$	$\mathbf{r}_3$
nal			
ass	$m_{12} = \dfrac{m_1 m_2}{m_1+m_2}$	$m_{3(12)} = \dfrac{m_3 M_{12}}{m_3+M_{12}}$	$M_{3(12)} = m_3+M_{12}$
oordinate	$\mathbf{r}_{12} = \mathbf{r}_1-\mathbf{r}_2$	$\mathbf{r}_{3(12)} = \mathbf{r}_3-\mathbf{R}_{12}$	$\mathbf{R}_{3(12)} = \dfrac{m_3\mathbf{r}_3+M_{12}\mathbf{R}_{12}}{M_3+M_{12}}$

The final coordinate transformation is thus

$$
\begin{aligned}
\mathbf{r}_{12} &= \mathbf{r}_1-\mathbf{r}_2, \\
\mathbf{r}_{3(12)} &= \mathbf{r}_3-\frac{m_1\mathbf{r}_1+m_2\mathbf{r}_2}{m_1+m_2}, \\
\mathbf{r}_g &= \mathbf{R}_{3(12)} = \frac{m_1\mathbf{r}_1+m_2\mathbf{r}_2+m_3\mathbf{r}_3}{M},
\end{aligned}
\tag{5.1.40}
$$

where M is the total mass. The Hamilton operator will be of the form

$$
\hat{H} = -\frac{\hbar^2}{2m_{12}}\Delta_{12}-\frac{\hbar^2}{2m_{3(12)}}\Delta_{3(12)}-\frac{\hbar^2}{2M}\Delta_g+\Phi, \tag{5.1.41}
$$

where

$$
m_{3(12)} = \frac{m_3(m_1+m_2)}{M} \tag{5.1.42}
$$

is the reduced mass of particle 3 relative to the (12) system. Since, for an isolated system, the potential Φ is independent of $\mathbf{r}_g$ one may again separate the centre-of-mass motion.

From the way in which the transformation (5.1.40) was generated in Table 1, it is seen that the total angular momentum of the system may be written as

$$\hat{\mathbf{J}} = \hat{\mathbf{L}}+\hat{\mathbf{S}}, \tag{5.1.43}$$

where $\hat{\mathbf{L}}$ is the angular momentum of the centre of mass, while $\hat{\mathbf{S}}$ is the intrinsic angular momentum. This may in turn be written as the sum of the angular momentum of particle 3 relative to the (12) system and the intrinsic angular momentum of the (12) system.

The above procedure can obviously be generalized to systems consisting of more than three particles.

Problems

5.1.1 Find the binding energy in the ground state for a μ-mesic hydrogen atom (proton and μ-meson), and find also the change in the binding energy if the nucleus is a deuteron. The nuclei are considered as point charges and the mass of the μ-meson is taken to be 207 times the electron mass. As a consequence of the above difference in binding energy, a μ-mesic hydrogen atom in hydrogen gas colliding with a normal deuterium atom, may transfer the μ-meson to the deuterium nucleus with release of energy. If this energy is transformed purely into kinetic energy of the two atoms, show that the proton receives an energy of about 50 eV.

5.1.2 Determine the energy spectrum for positronium (a bound system consisting of an electron and a positron) and compare for the ground state the size of the positronium system and the hydrogen atom.

5.1.3 Assume that the attraction between a neutron and a proton by nuclear forces can be represented by a spherical potential well with a radius $a = 2\times 10^{-13}$ cm. Give an approximate estimate of the depth of the well when the ground state of the system ($l = 0$) represents the deuterium nucleus, whose binding energy is $E_D = 2{\cdot}2$ MeV [compare problem 4.5.2].

5.1.4 For molecules at low excitation energies one need not consider the system of electrons in detail. They simply determine the forces that bind together the nuclei in the molecule. Show that, if a diatomic molecule can be considered as a rigid system capable of rotation only, the energy spectrum is of the form

$$E_l = \frac{\hbar^2 l(l+1)}{2I}, \qquad (l = 0, 1, 2, \ldots\ldots),$$

where I is the moment of inertia of the molecule with respect to a line through the centre of mass perpendicular to, and bisecting, the line joining the two nuclei.

If the system is not completely rigid one may, to first order, describe the binding by assuming that the nuclei are bound to the above considered equilibrium distance r_0 by a force $k(r-r_0)$.

By means of a series expansion about r_0 of the effective potential in the relative motion, show that the total energy spectrum for small excitation energies is of the form

$$E_{l,n} = \frac{\hbar^2 l(l+1)}{2I} + (\tfrac{1}{2}+n)\hbar\left(\frac{k}{m_0}\right)^{\frac{1}{2}} + \text{constant},$$

$$l = 0, 1, 2, \ldots, \quad n = 0, 1, 2, \ldots.$$

5.2 PARTICLE SYSTEMS IN EXTERNAL FIELDS

The separation of the centre of mass motion from the intrinsic motion of a system of particles was discussed above for the case of an isolated particle system. If there are external forces on the system, an exact separation can only be achieved if the forces correspond to a homogenous gravitational field. However, an approximate separation is obtained if the external forces are weak compared to the forces which act between the particles. Under these circumstances the Hamiltonian can be written in the form

$$\hat{H} = \hat{H}_{\text{int}} + \hat{H}_g, \tag{5.2.1}$$

where

$$\hat{H}_g = -\frac{\hbar^2}{2M}\Delta_g + \Phi(\mathbf{r}_g). \tag{5.2.2}$$

The solution to the Schrödinger equation (5.1.8) is then

$$\psi = \psi_g(\mathbf{r}_g, t)\cdot\chi \exp(-\mathrm{i}E_{\text{int}}t/\hbar), \tag{5.2.3}$$

where

$$\hat{H}_g\psi_g = \mathrm{i}\hbar\frac{\partial}{\partial t}\psi_g, \tag{5.2.4}$$

and where χ is an eigenfunction of the Hamiltonian $\hat{H}_{\text{int}}$:

$$\hat{H}_{\text{int}}\chi = E_{\text{int}}\chi. \tag{5.2.5}$$

In such cases the system as a whole behaves completely as a particle. The intrinsic wave function will, in general, represent a state of definite parity, angular momentum, etc., but it is seen from (5.2.2) and (5.2.4) that for the determination of the centre of mass motion only the total mass is of importance. In the preceding chapters we have already utilized this result when we considered the atomic nucleus in He^+, Li^{2+}, etc. as a point charge or even as a static charge distribution. In many atomic systems such a complete separation is not possible, mainly because the particle system possesses a magnetic moment which interacts with the surroundings. New terms will appear in the Hamiltonian, of the form $V(\hat{\mathbf{S}}, \mathbf{r}_g)$, where $\hat{\mathbf{S}}$ is the operator for the total angular momentum of the particle system in the centre of mass system.

As we shall discuss in the remaining part of this section, one may, even in the presence of such terms, perform a separation of the centre of mass motion. One must, however, then specify the value of the total angular momentum, which we call $\hbar\sqrt{s(s+1)}$ and its component, $m_s\hbar$, along the z-axis.

Let us denote the intrinsic wave function thus specified by $\chi(s, m_s)$, which means that χ, besides satisfying (5.2.5), also satisfies the equations

$$\hat{S}^2\chi(s, m_s) = \hbar^2 s(s+1)\chi(s, m_s) \tag{5.2.6}$$

and

$$\hat{S}_z\chi(s, m_s) = \hbar m_s\chi(s, m_s), \tag{5.2.7}$$

and that the state is thus uniquely determined.

It is now essential for the following that, even though the operators $\hat{S}_x$, $\hat{S}_y$ and $\hat{S}_z$ depend explicitly on the intrinsic coordinates, their action on an intrinsic state simply consists of a rotation of the intrinsic angular momentum. This is most easily seen by using the commutation relations

$$[\hat{S}_x, \hat{S}_y] = \mathrm{i}\hbar\hat{S}_z, \tag{5.2.8}$$

$$[\hat{S}_y, \hat{S}_z] = \mathrm{i}\hbar\hat{S}_x, \tag{5.2.9}$$

$$[\hat{S}_z, \hat{S}_x] = \mathrm{i}\hbar\hat{S}_y, \tag{5.2.10}$$

for the operators $\hat{S}_x$, $\hat{S}_y$ and $\hat{S}_z$. These relations are consequences of (4.2.11) since $\hat{\mathbf{S}}$ is a sum of the angular momenta $\hat{\mathbf{L}}_i$ of the reduced particles.

Multiplying equation (5.2.9) by i and subtracting it from (5.2.10), one obtains

$$\hat{S}_z\hat{S}_+ - \hat{S}_+\hat{S}_z = \hbar\hat{S}_+, \tag{5.2.11}$$

where

$$\hat{S}_+ = \hat{S}_x + \mathrm{i}\hat{S}_y. \tag{5.2.12}$$

Correspondingly, one finds by addition

$$\hat{S}_z\hat{S}_- - \hat{S}_-\hat{S}_z = -\hbar\hat{S}_-, \tag{5.2.13}$$

with

$$\hat{S}_- = \hat{S}_x - \mathrm{i}\hat{S}_y. \tag{5.2.14}$$

Letting the operators on both sides of equation (5.2.11) operate on the state χ, one finds from (5.2.7)

$$\hat{S}_z[\hat{S}_+\chi(s, m_s)] = \hbar(m_s+1)\,[\hat{S}_+\chi(s, m_s)]. \tag{5.2.15}$$

This equation shows that the state $\hat{S}_+\chi(s, m_s)$ is an eigenstate of $\hat{S}_z$ with the eigenvalue $\hbar(m_s+1)$. Since $\hat{S}_+$ and $\hat{H}_{\text{int}}$ commute, the state is furthermore an eigenfunction of $\hat{H}_{\text{int}}$ with energy E_{int}, and since $\hat{S}_+$ and $\hat{S}^2$ commute, the state remains an eigenfunction for $\hat{S}^2$ corresponding to the eigenvalue $\hbar^2 s(s+1)$. From the assumption of the uniqueness of the solutions of (5.2.5), (5.2.6) and (5.2.7), it follows finally that $\hat{S}_+\chi(s, m_s)$ must be proportional to $\chi(s, m_s+1)$, i.e.

$$\hat{S}_+\chi(s, m_s) = A(s, m_s)\chi(s, m_s+1), \tag{5.2.16}$$

where A is a number.

In a completely analogous way, one finds from (5.2.13) the relation

$$\hat{S}_-\chi(s, m_s) = B(s, m_s)\chi(s, m_s-1), \tag{5.2.17}$$

where B is a number.

We shall now try to determine the constants A and B which are expected to depend on s and m_s. To this end we use the fact that $\hat{S}_+$ and $\hat{S}_-$, according to (5.2.8) and the definitions (5.2.12) and (5.2.13), satisfy the equation

$$\hat{S}_+\hat{S}_- = \hat{S}^2 - \hat{S}_z^{\,2} + \hbar\hat{S}_z. \tag{5.2.18}$$

Letting both sides of this operator equation act on $\chi(s, m_s+1)$ one obtains

$$B(s, m_s+1)A(s, m_s) = \hbar^2[s(s+1) - (m_s+1)^2 + m_s + 1]. \tag{5.2.19}$$

We now multiply equation (5.2.16) by $\chi^*(s, m_s+1)$ and integrate over all intrinsic degrees of freedom. Since $\chi(s, m_s)$ is assumed to be normalized one obtains

$$\int \chi^*(s, m_s+1)\hat{S}_+\chi(s, m_s)\mathrm{d}\tau_{\text{int}} = A(s, m_s). \tag{5.2.20}$$

Since $\hat{\mathbf{S}}$ is a Hermitian operator, the left-hand side of this equation can be transformed into

$$\int [\hat{S}_-\chi(s, m_s+1)]^*\chi(s, m_s)\mathrm{d}\tau_{\mathrm{int}} \tag{5.2.21}$$

which, according to (5.2.17), is equal to $B^*(s, m_s+1)$, and we have thus shown that

$$A(s, m_s) = B^*(s, m_s+1). \tag{5.2.22}$$

Inserting this into (5.2.19) we finally obtain

$$|A(s, m_s)|^2 = \hbar^2(s+m_s+1)(s-m_s) \tag{5.2.23}$$

or

$$A(s, m_s) = \hbar\sqrt{(s+m_s+1)(s-m_s)}. \tag{5.2.24}$$

Strictly speaking, one can only conclude that A is given by (5.2.24) apart from a factor exp (iα), where α is a real constant. This phase factor is, in principle, unknown, since one can always multiply wave functions $\chi(s, m_s)$ for different states by arbitrary phase factors. For the special eigenfunctions of the intrinsic angular momentum of two particles, $\chi(s, m_s) = R(r)\cdot Y_{s,m_s}(\theta, \phi)$, which we considered in Chapter 4, the expression (5.2.24) holds without phase factors and it has become a common convention to normalize eigenstates of many-particle systems in the same fashion.

We have now concluded the investigation of how the angular momentum operators

$$\hat{S}_x = \tfrac{1}{2}(\hat{S}_+ + \hat{S}_-)$$

$$\hat{S}_y = \tfrac{1}{2}\mathrm{i}(\hat{S}_+ - \hat{S}_-) \tag{5.2.25}$$

and $\hat{S}_z$

act on the intrinsic states, and we have seen how their action on a wave function $\chi(s, m_s)$ can be expressed by a linear combination of eigenfunctions belonging to other values of m_s but to the same value of s. This implies that the intrinsic motion can also be separated from the centre of mass motion in cases where the Hamiltonian contains an interaction between the total intrinsic angular momentum and external fields.

5.3 SPIN

The separation of the centre of mass motion from the intrinsic motion is of great importance in atomic physics. One encounters many atomic systems where the binding forces between the parts of the system are much larger than the external forces. If the system possesses an intrinsic angular momentum it should be appreciated that even weak forces can rotate the angular momentum, for instance by interacting with the magnetic dipole moment of the system. As we have seen, however, even in such cases one can eliminate the intrinsic motion and find the behaviour of the system as a whole merely from a knowledge of the total mass M and the magnitude of the intrinsic angular momentum s. The two quantities M and s can only be calculated if we know the intrinsic structure. One would expect from our knowledge of the two-body problem that the quantities s and m_s can only be integers. In the calculations which we performed in 5.2 this was more or less understood although it was not used directly anywhere, and it is therefore of interest to investigate whether the formalism might allow other values of s and m_s.

In this connexion it is noted first that the equations (5.2.16) and (5.2.17) show that if the system can exist in a state with eigenvalue $m\hbar$ for $\hat{S}_z$, then there exist also states with the eigenvalues

$$m_s\hbar = (m+1)\hbar,\ (m+2)\hbar, \text{ etc.}$$

and

$$m_s\hbar = (m-1)\hbar,\ (m-2)\hbar, \text{ etc.} \tag{5.3.1}$$

This conclusion remains valid as long as the coefficients A and B do not vanish. Now, however, one cannot allow arbitrarily large values for $|m_s|$ since the quantity $|A|^2$ in (5.2.23) would then become negative. We conclude from this that the sequence of numbers, $m, m+1, m+2, \ldots.$ must break off at an upper limit $m_s = m_{\max}$, and that the sequence $m-1, m-2$, must break off at a lower limit $m_s = m_{\min}$. From these two conditions we conclude that $A(s, m_{\max}) = 0$ and also $B(s, m_{\min}) = 0$, i.e. from (5.2.22) and (5.2.24)

$$m_{\max} = s \qquad \text{and} \qquad m_{\min} = -s. \tag{5.3.2}$$

Since, however, the numbers $m_{\max}$ and $m_{\min}$, must according to (5.3.1), differ by an integer we find that

$$m_{\max} - m_{\min} = 2s = \text{integer}. \tag{5.3.3}$$

Besides the integer solutions for s we see that our formalism also allows half-integer values of s. It is well known that one has found in nature a number of particles with half-integer spin. This is of course not inconsistent with our previous calculations on angular momenta; it only shows that the intrinsic structure of particles with half-integer spin cannot be described as a non-relativistic motion of spinless particles.

The stationary states for systems with intrinsic angular momentum s can be written, as we have seen, in the form

$$u = \sum_{m=-s}^{m=s} u_m(\mathbf{r})\chi(s, m), \tag{5.3.4}$$

where we have omitted the index g on the centre of mass coordinate. If with this wave function one wants to evaluate the expectation value of a physical quantity $F(\mathbf{r}, \hat{\mathbf{P}})$ which only concerns the centre of mass motion one finds

$$\begin{aligned}\langle F(\mathbf{r}, \hat{\mathbf{P}})\rangle &\\ &= \sum_{m, m'} \int \mathrm{d}^3 r \int \mathrm{d}\tau_{\text{int}} \chi^*(s, m)\chi(s,m')u_m^*(\mathbf{r})F(\mathbf{r},\hat{\mathbf{P}})u_{m'}(\mathbf{r}) \\ &= \sum_m \int \mathrm{d}^3 r u_m^*(\mathbf{r})F(\mathbf{r},\hat{\mathbf{P}})u_m(\mathbf{r}).\end{aligned} \tag{5.3.5}$$

We have here first integrated over the intrinsic coordinates and used the orthogonality of the states $\chi(s, m)$. The result (5.3.5) shows that the only property of the intrinsic wave functions which we need in (5.3.4) is that they form an orthonormal system. One therefore often uses a notation in which the wave functions χ are left out. The wave function u is then considered as a column vector

$$u = \begin{Bmatrix} u_s\ (\mathbf{r}) \\ u_{s-1}(\mathbf{r}) \\ \vdots \\ \vdots \\ u_{-s}\ (\mathbf{r}) \end{Bmatrix}. \tag{5.3.6}$$

A notation of this type is of course equivalent to substituting the intrinsic wave functions by the column vectors

$$\chi(s, s) = \begin{Bmatrix} 1 \\ 0 \\ 0 \\ \vdots \\ 0 \end{Bmatrix}, \chi(s, s-1) = \begin{Bmatrix} 0 \\ 1 \\ 0 \\ \vdots \\ 0 \end{Bmatrix}, \ldots, \chi(s, -s) = \begin{Bmatrix} 0 \\ 0 \\ \vdots \\ 1 \end{Bmatrix} \tag{5.3.7}$$

and the orthogonality of the functions $\chi(s, m)$ is then expressed by the fact that the $(2s+1)$-dimensional vectors in (5.3.7) are orthonormal in the usual sense of vector algebra.

The operators $\hat{\mathbf{S}}$ lead, as we have seen, to linear transformations between the $2s+1$ vectors (5.3.7), and can therefore be expressed by quadratic matrices of rank $2s+1$. We find from (5.2.16), (5.2.17) and (5.2.22)

$$\hat{S}_+ = \begin{Bmatrix} 0 & A(s, s-1) & 0 \ldots\ldots\ldots\ldots 0 \\ 0 & 0 & A(s, s-2) \ldots\ldots 0 \\ 0 & 0 & 0 \ldots\ldots\ldots\ldots 0 \\ \ldots\ldots & \ldots\ldots & \ldots\ldots A(s, -s) \\ 0 & 0 & 0 \ldots\ldots\ldots\ldots 0 \end{Bmatrix} \tag{5.3.8}$$

and

$$\hat{S}_- = \begin{Bmatrix} 0 & 0 \ldots\ldots 0 & \ldots\ldots\ldots & 0 \\ A^*(s, s-1) & 0 & 0 \ldots\ldots\ldots & 0 \\ 0 & A^*(s, s-2) & 0 \ldots\ldots\ldots & 0 \\ \ldots\ldots & \ldots\ldots & \ldots\ldots & \\ 0 & 0 & 0 \ldots A^*(s, -s) & 0 \end{Bmatrix} \tag{5.3.9}$$

as well as from (5.2.7)

$$\hat{S}_z = \hbar \begin{Bmatrix} s & 0 & 0 & 0 \ldots\ldots\ldots 0 \\ 0 & s-1 & 0 & 0 \ldots\ldots\ldots 0 \\ 0 & 0 & s-2 & 0 \ldots\ldots\ldots 0 \\ \ldots\ldots & \ldots\ldots & \ldots\ldots & \ldots\ldots \\ 0 & 0 & 0 & 0 \ldots\ldots -s \end{Bmatrix}. \tag{5.3.10}$$

From these expressions one finds easily through the definition (5.2.25) the matrices for $\hat{S}_x$ and $\hat{S}_y$. For the special case of $s = \frac{1}{2}$ one finds

$$\hat{\mathbf{S}} = \tfrac{1}{2}\hbar\boldsymbol{\sigma} \tag{5.3.11}$$

where

$$\sigma_x = \begin{Bmatrix} 0 & 1 \\ 1 & 0 \end{Bmatrix}, \quad \sigma_y = \begin{Bmatrix} 0 & -\mathrm{i} \\ \mathrm{i} & 0 \end{Bmatrix}, \quad \sigma_z = \begin{Bmatrix} 1 & 0 \\ 0 & -1 \end{Bmatrix}. \tag{5.3.12}$$

The matrices σ_x, σ_y and σ_z are often called the Pauli spin matrices.

Problem

5.3.1 Show that the spin matrices for $s = 1$ may be written

$$\hat{S}_x = \frac{\hbar}{\sqrt{2}}\begin{Bmatrix} 0 & 1 & 0 \\ 1 & 0 & 1 \\ 0 & 1 & 0 \end{Bmatrix}$$

$$\hat{S}_y = \frac{\hbar}{\sqrt{2}}\begin{Bmatrix} 0 & -\mathrm{i} & 0 \\ \mathrm{i} & 0 & \mathrm{i} \\ 0 & -\mathrm{i} & 0 \end{Bmatrix}$$

$$\hat{S}_z = \hbar\begin{Bmatrix} 1 & 0 & 0 \\ 0 & 0 & 0 \\ 0 & 0 & -1 \end{Bmatrix}$$

5.4 HARTREE TREATMENT OF MANY-BODY PROBLEMS

In the remaining part of this chapter we shall discuss some important features of the intrinsic structure of systems containing many particles. The accurate description of such systems is, in general, extremely complicated, and in almost all applications one is forced to use quite radical approximation methods. We are usually concerned with systems of electrons or nucleons where the Pauli exclusion principle is of decisive importance for the structure. We shall, however, postpone the detailed introduction of the Pauli principle until the basic approximations are formulated.

The most important approximation method for the treatment of many-particle systems is the so-called Hartree approximation, in which one approximates the true wave function of the N-particle system by a product wave function of the type (5.1.13). We consider the stationary Schrödinger equation for the N-particle system

$$\hat{H}u(\mathbf{r}_1,\mathbf{r}_2,\ldots\mathbf{r}_N) = Eu(\mathbf{r}_1,\mathbf{r}_2,\ldots\mathbf{r}_N), \tag{5.4.1}$$

and write the Hamilton operator in the form

$$\hat{H} = \sum_{i=1}^{N} \hat{H}_i + V(\mathbf{r}_1, \mathbf{r}_2, \ldots \mathbf{r}_N), \tag{5.4.2}$$

where

$$\hat{H}_i = -\frac{\hbar^2}{2m_i}\Delta_i + \Phi_i(\mathbf{r}_i)$$

is that part of the Hamilton operator which only depends on the degrees of freedom of the ith particle, while V describes the interaction between the particles. We try to find the 'best' approximate solution of (5.4.1) of the form

$$u(\mathbf{r}_1, \ldots \mathbf{r}_N) = u_1(\mathbf{r}_1)u_2(\mathbf{r}_2) \cdots u_N(\mathbf{r}_N) \tag{5.4.3}$$

by means of the variational principle discussed in 2.6. It is easily seen to be applicable also for problems of several particles. To find an approximate solution of (5.4.1) we thus make independent variations of $u_1, u_2 \ldots u_N$, and demand that

$$\langle E \rangle = \frac{\int u\hat{H}u\,\mathrm{d}\tau_1\mathrm{d}\tau_2 \cdots \mathrm{d}\tau_N}{\int (u)^2\mathrm{d}\tau_1 \cdots \mathrm{d}\tau_N} \tag{5.4.4}$$

be minimal, where we have assumed the wave functions to be real. Introducing (5.4.2) and (5.4.3) into (5.4.4), we find

$$\langle E \rangle = \sum_{i=1}^{N} \frac{\int u_i\hat{H}_iu_i\mathrm{d}\tau_i}{\int (u_i)^2\mathrm{d}\tau_i} + \frac{\int uVu\,\mathrm{d}\tau_1 \cdots \mathrm{d}\tau_N}{\int (u)^2\mathrm{d}\tau_1 \cdots \mathrm{d}\tau_N}, \tag{5.4.5}$$

where each term in the sum only contains the wave function $u_i(\mathbf{r}_i)$ of the ith particle.

We now perform a variation of the wave function of particle k, and obtain in analogy to (2.6.2):

$$\begin{aligned}\delta_k\langle E \rangle = {} & 2\frac{\int \delta u_k(\hat{H}_k - \int u_k\hat{H}_ku_k\mathrm{d}\tau_k)u_k\mathrm{d}\tau_k}{\int (u_k)^2\mathrm{d}\tau_k} \\ & +2\frac{\int \delta u_k \int u^{(k)}Vu^{(k)}\mathrm{d}\tau^{(k)}}{\int (u)^2\mathrm{d}\tau_1 \cdots \mathrm{d}\tau_N} \\ & -2\frac{\int \delta u_ku_k\mathrm{d}\tau_k \int (u^{(k)})^2\mathrm{d}\tau^{(k)} \int uVu\,\mathrm{d}\tau_1 \cdots \mathrm{d}\tau_N}{(\int (u)^2\mathrm{d}\tau_1 \cdots \mathrm{d}\tau_N)^2}.\end{aligned} \tag{5.4.6}$$

We have here introduced the notation

$$u^{(k)} = u_1(\mathbf{r}_1)u_2(\mathbf{r}_2) \cdots u_{k-1}(\mathbf{r}_{k-1})u_{k+1}(\mathbf{r}_{k+1}) \cdots u_N(\mathbf{r}_N) \qquad (5.4.7)$$

for the wave function (5.4.3) in which the factor $u_k(\mathbf{r}_k)$ has been left out, and $\mathrm{d}\tau^{(k)}$ denotes the corresponding volume element

$$\mathrm{d}\tau_1\mathrm{d}\tau_2 \cdots \mathrm{d}\tau_{k-1}\mathrm{d}\tau_{k+1} \cdots \mathrm{d}\tau_N.$$

Utilizing

$$\int (u)^2\mathrm{d}\tau_1 \cdots \mathrm{d}\tau_N = \int (u_k)^2\mathrm{d}\tau_k \int (u^{(k)})^2\mathrm{d}\tau^{(k)}, \qquad (5.4.8)$$

we may write (5.4.6) in the form

$$\delta_k\langle E\rangle = 2\frac{\int \delta u_k[\hat{H}_k - \langle H_k\rangle + \bar{V}_k - \langle \bar{V}_k\rangle]u_k\mathrm{d}\tau_k}{\int (u_k)^2\mathrm{d}\tau_k}, \qquad (5.4.9)$$

where $\bar{V}_k$ denotes the average of $V(\mathbf{r}_1,\mathbf{r}_2, \ldots \mathbf{r}_N)$ over all coordinates except $\mathbf{r}_k$, i.e.

$$\bar{V}_k = \bar{V}_k(\mathbf{r}_k) = \frac{\int u^{(k)}V(\mathbf{r}_1, \ldots \mathbf{r}_N)u^{(k)}\mathrm{d}\tau^{(k)}}{\int (u^{(k)})^2\mathrm{d}\tau^{(k)}}. \qquad (5.4.10)$$

The expectation values in (5.4.9) need only be taken over the wave function $u_k(\mathbf{r}_k)$.

It is now seen that, since (5.4.9) should vanish for all variations δu_k, the wave function u_k must satisfy the equation

$$[\hat{H}_k + \bar{V}_k(\mathbf{r}_k)]\, u_k(\mathbf{r}_k) = E_k u_k(\mathbf{r}_k), \qquad (5.4.11)$$

where E_k is given by

$$\begin{aligned} E_k &= \langle H_k\rangle + \langle \bar{V}_k\rangle \\ &= \langle H_k\rangle + \langle V\rangle. \end{aligned} \qquad (5.4.12)$$

Performing independent variations of all the other single particle wave functions, one sees that equation (5.4.11) must hold for all N particles. These are the Hartree equations. They are formally of the type of single particle wave equations which, however, are coupled through the potential $\bar{V}_k(\mathbf{r}_k)$, which depends on the wave functions of all the other particles. We thus have, according to (5.4.10),

$$\bar{V}_k(\mathbf{r}_k) = \int V(\mathbf{r}_1 \ldots \mathbf{r}_N)P_1(\mathbf{r}_1) \cdots P_{k-1}(\mathbf{r}_{k-1})P_{k+1}(\mathbf{r}_{k+1}) \cdots P_N(\mathbf{r}_N) \cdot \mathrm{d}\tau_1\mathrm{d}\tau_2 \cdots \mathrm{d}\tau_{k-1}\mathrm{d}\tau_{k+1} \cdots \mathrm{d}\tau_N, \qquad (5.4.13)$$

where $P_i(\mathbf{r}_i)$ is the probability density $|u_i(\mathbf{r}_i)|^2/\int |u_i(\mathbf{r}_i)|^2\mathrm{d}\tau_i$ of the ith particle. It should be noted that although we have derived the result (5.4.11) under the assumption of real wave functions, the results (5.4.11)–(5.4.13) hold also for complex wave functions.

The Hartree equations have the intuitive physical significance that the best independent particle description is obtained if any one of the particles move in the average field (5.4.13) of all the others. The fact that we do not have a pure independent particle description shows up through the fact that the total energy (5.4.5) is given by

$$\langle E\rangle = \sum_{i=1}^{N} \langle H_i\rangle + \langle V\rangle, \tag{5.4.14}$$

which in general differs from the sum of the single particle energies

$$\sum_{i=1}^{N} E_i = \sum_{i=1}^{N} (\langle H_i\rangle + \langle V\rangle). \tag{5.4.15}$$

A solution of the Hartree equations can be obtained by iteration. In the first place, one must guess reasonable values $\bar{V}_k^{(0)}$ for the average potentials. One then solves the Schrödinger equations (5.4.11) and finds a solution

$$u^{(0)} = u_1^{(0)}(\mathbf{r}_1)u_2^{(0)}(\mathbf{r}_2) \cdots\cdots\cdots u_N^{(0)}(\mathbf{r}_N), \tag{5.4.16}$$

where

$$\left[\frac{\hat{\mathbf{p}}_i^{\,2}}{2m_i} + V_i^{(0)}(\mathbf{r}_i)\right] u_i^{(0)}(\mathbf{r}_i) = E_i^{(0)} u_i^{(0)}(\mathbf{r}_i). \tag{5.4.17}$$

With this solution one can now evaluate improved values for the average potential $\bar{V}_k^{(1)}$, setting

$$\bar{V}_k^{(1)}(\mathbf{r}_1) = \int [u_{(k)}^{(0)}]^* V(\mathbf{r}_1, \ldots \mathbf{r}_N)\, u_{(k)}^{(0)}\, \mathrm{d}\tau^{(k)}, \tag{5.4.18}$$

where the average value is taken over the coordinates of all particles except the kth. With the improved potentials one may then evaluate improved wave functions $u^{(1)}$, and the process can be repeated until one has found a 'self-consistent' set of potentials and associated wave functions.

The wave functions which have been determined by this procedure are not, of course, solutions of the exact Schrödinger equation of the system. In contrast to what is to be expected, they do not contain any correlation between the positions of the particles. Nevertheless, the Hartree model gives a rather good description of atomic systems and is especially important for the classification of the stationary states.

In the following we shall mainly discuss many-body problems on the basis of the Hartree description; we consider first its application to the description of heavy atoms. We consider here the motion of N electrons (of mass m) in the Coulomb field of a nucleus with charge Ze. We assume this to be infinitely heavy, and the Hamiltonian for the atom is therefore

$$\hat{H} = \sum_{i=1}^{N}\left(\frac{\hat{\mathbf{p}}_i^2}{2m} - \frac{Ze^2}{|\mathbf{r}_i|}\right) + \sum_{i>j}\frac{e^2}{|\mathbf{r}_i - \mathbf{r}_j|}. \tag{5.4.19}$$

The position vector $\mathbf{r}_i$ of the ith electron is here measured from the atomic nucleus, which is assumed to be a point charge.

A first approximation in a Hartree treatment would be to neglect the last term in (5.4.19), which represents the mutual Coulomb repulsion between the electrons. In this approximation the wave functions for the stationary states are products of the eigenfunctions u_{nlm} for a hydrogen-like atom with nuclear charge Ze [cf. equation (4.3.22)]:

$$u^{(0)} = u_{n_1 l_1 m_1}(\mathbf{r}_1) \cdot u_{n_2 l_2 m_2}(\mathbf{r}_2) \cdot \cdots \cdot u_{n_N l_N m_N}(\mathbf{r}_N), \tag{5.4.20}$$

where one should take into account the restriction imposed by the Pauli exclusion principle, i.e. in each state n, l, m there can be at most two electrons. Since the energy E_i is independent of the magnetic quantum numbers m_i, the energy levels of the atom are given by the main quantum number n_i and the angular momentum quantum number l_i for all the electrons. Instead of l one often uses the spectroscopic notation in which a definite atomic energy level is denoted by the symbol

$$(1s)^\alpha\,(2s)^\beta\,(2p)^\gamma\,(3s)^\delta \ldots\ldots,$$

meaning that α electrons are in the state with $n = 1$ and $l = 0$, β electrons are in the state with $n = 2$ and $l = 0$, etc.

A better approximation for the atomic wave functions is obtained by the next step in the Hartree approximation, where one uses the wave function (5.4.20) with a given set of quantum numbers to determine the average value of the last term in (5.4.19). We will not perform these calculations here but merely indicate a first approximation to a systematic improvement of the wave functions. In order to do this we consider the illustration Fig. 4.5 of the radial probability distribution of electrons in hydrogen-like atoms. It is seen here that an electron in a $1s$-state is mainly to be found in the atom inside the electrons in states with $n = 2$ and that the electrons in $n = 2$ orbits are found almost completely inside the electrons with $n = 3$, etc. This is the justification for the use of the concept of electron shells in atoms. If we consider an electron in a state with $n = n_i$, one may from this shell structure conclude [cf. problem 5.4.1] that all electrons in states with $n < n_i$ will give approximately the contribution e^2/r^2 to the average force while electrons with $n > n_i$ will give a vanishing contribution. The contribution to the average force from other electrons in the same shell ($n = n_i$) may be shown to be approximately $0{\cdot}3\ e^2/r^2$ [cf. problem 5.4.2]. Thus the result of this rough estimate is that the magnitude of the force on the ith electron is

$$F_i(\mathbf{r}_i) = \frac{Z^{(i)}{}_{eff}e^2}{r_i^2}, \tag{5.4.21}$$

with

$$Z_{eff}^{(i)} = Z - \sum_{n<n_i} N_n - \tfrac{1}{3}(N_{n_i} - 1), \tag{5.4.22}$$

where N_n is the number of electrons in the state with main quantum number n.

The improved wave functions $u^{(1)}$ are again products of hydrogen-like eigenfunctions (5.4.20), where the wave function for each electron is evaluated, however, with different effective nuclear charge. This primitive Hartree model, possibly with a somewhat more refined prescription for the calculation of Z_{eff}, is called the Slater model. The Slater model gives a reasonable description of many properties of the atom.

Problems

5.4.1 Use the explicit expression (4.1.21) to show that the average electric potential $\bar{V}$ from three electrons in the $l = 1$ states, with $m = 1, 0$ and -1 respectively, is spherically symmetric.

5.4.2 Consider the ground state $(1s)^2$ of helium in the Hartree model, where the wave function is assumed to be the product of two hydrogen-like $1s$ states [cf. (4.3.16)]:

$$u(\mathbf{r}_1,\mathbf{r}_2) = 2\left(\frac{Z}{a_0}\right)^3 \exp\left(-\frac{Zr_1}{a_0}\right) \exp\left(-\frac{Zr_2}{a_0}\right).$$

Allowing Z to be a free parameter, show by the variational principle that the best wave function of this type is obtained for

$$Z = Z_{\text{eff}} = 2 - \frac{5}{16}.$$

5.5 THE PAULI PRINCIPLE

The stationary states for an atom in the Slater model, although less degenerate than the above primitive ansatz, still possess a high degree of degeneracy. This degeneracy is partly removed if one carries through the next stage in the Hartree approximation, taking into account that states with a definite magnetic quantum number give rise to an average field which is not spherically symmetric (see 5.6). Furthermore, electrons are known to have spin $s = \frac{1}{2}$. Any state described by quantum numbers (n, l, m) is therefore doubly degenerate since the spatial wave function u_{nlm} should be multiplied by the spin wave function $\chi_{+\frac{1}{2}}$ corresponding to $m_s = \frac{1}{2}$, or by $\chi_{-\frac{1}{2}}$ corresponding to $m_s = -\frac{1}{2}$. If one considers in the Hamiltonian the interaction between the intrinsic magnetic moment of the electrons and the magnetic field which is created by their motion, this degeneracy will also be removed.

There is, however, one type of degeneracy which is not removed. This is the degeneracy which is due to the fact that the state

$$u = u_{n_1 l_1 m_1}(\mathbf{r}_1)\, u_{n_2 l_2 m_2}(\mathbf{r}_2) \dots \tag{5.5.1}$$

has the same energy as the state

$$u = u_{n_2 l_2 m_2}(\mathbf{r}_1)\, u_{n_1 l_1 m_1}(\mathbf{r}_2) \dots \tag{5.5.2}$$

which is obtained by exchanging the coordinates for the electrons 1 and 2. The equality of the energies corresponding to the states (5.5.1) and (5.5.2) is a consequence of the fact that the Hamilton operator (5.4.19) commutes with the operator $\hat{P}_{ik}$ which interchanges the coordinates of two arbitrary electrons i and k. From this fact we conclude also that the operators $\hat{P}_{ik}$ are constants of the motion, and the eigenstates of $\hat{H}$ are therefore also eigenstates of $\hat{P}_{ik}$. We may easily show that the eigenvalue equation

$$\hat{P}_{ik} u = P_{ik} u, \tag{5.5.3}$$

where P_{ik} is a number, can only have the eigenvalues $P_{ik} = \pm 1$. Thus we obviously have

$$(\hat{P}_{ik})^2 = 1. \tag{5.5.4}$$

i.e.

$$(\hat{P}_{ik})^2 u = (P_{ik})^2 u = u. \tag{5.5.5}$$

When we look for eigenfunctions of the Hamilton operator (5.4.19) we should thus look for them among functions which are either symmetric, $P_{ik} = 1$, or antisymmetric, $P_{ik} = -1$, under exchange of two arbitrary particles. We are here in a situation like the one described in 3.5, where a good approximation to the eigenfunctions is only obtained by using linear combinations of (3.5.2) and (3.5.5) such that the wave functions are simultaneously eigenfunctions of the parity operator.

If we limit ourselves to the two-electron problem (e.g. helium atom) we should, rather than using the wave function

$$u = u_{n_1 l_1 m_1}(\mathbf{r}_1)\, u_{n_2 l_2 m_2}(\mathbf{r}_2), \tag{5.5.6}$$

use either the symmetric combination

$$u_+ = \frac{1}{\sqrt{2}}\,[u_{n_1 l_1 m_1}(\mathbf{r}_1)\, u_{n_2 l_2 m_2}(\mathbf{r}_2) + u_{n_2 l_2 m_2}(\mathbf{r}_1)\, u_{n_1 l_1 m_1}(\mathbf{r}_2)], \tag{5.5.7}$$

or the antisymmetric combination

$$u_- = \frac{1}{\sqrt{2}}\,[u_{n_1 l_1 m_1}(\mathbf{r}_1)\, u_{n_2 l_2 m_2}(\mathbf{r}_2) - u_{n_2 l_2 m_2}(\mathbf{r}_1)\, u_{n_1 l_1 m_1}(\mathbf{r}_2)]. \tag{5.5.8}$$

We have here introduced the factor $1/\sqrt{2}$ which ensures that u_+ and u_- are normalized if u_{nlm} is normalized. If, however, $n_1 = n_2$, $l_1 = l_2$ and $m_1 = m_2$, there exists, as one sees, only one (symmetric) wave function

$$u_+ = u_{n_1 l_1 m_1}(\mathbf{r}_1)\, u_{n_1 l_1 m_1}(\mathbf{r}_2). \tag{5.5.9}$$

The wave functions (5.5.7) and (5.5.8) contain, in contrast to (5.5.6), correlations between the positions of the electrons. Thus it is seen that u_- vanishes for $\mathbf{r}_1 = \mathbf{r}_2$. This implies also that the expectation value for the energy of a helium atom in a state described by u_- is, in general, lower than the energy of the corresponding state u_+, since the interaction energy

$$\frac{e^2}{|\mathbf{r}_1 - \mathbf{r}_2|}$$

in the Hamilton operator (5.4.19) gives a large contribution when $\mathbf{r}_1 \approx \mathbf{r}_2$. Now it is known from atomic physics that the bound states in the helium atom are described quite well by the configurations

$(1s)(nl)$ where one electron is in the $1s$ state. It is also known, however, that the level scheme actually consists of two sets of states with almost the same energy, namely the states for ortho- and para-helium. The only state which is not split is the ground state, which only appears in the para-helium spectrum. This is in complete accordance with what we should expect from the above if we identify the symmetric states with para-helium and the antisymmetric ones with ortho-helium.

It is also a consequence of this identification that transitions between ortho- and para-helium under emission of electromagnetic waves is forbidden. The interaction with external electromagnetic fields giving rise to transitions between stationary states must also be symmetric in the coordinates of the two electrons, and the quantum number P_{12} is therefore also a constant of the motion under such processes.

The above considerations are only strictly correct as long as we neglect the weak forces which may influence the spin of the electrons. If we take these forces into account the Hamiltonian no longer commutes with the operator $\hat{P}_{ik}$ which interchanges the coordinates of two electrons. One may then, however, introduce a new operator $\hat{P}'_{ik}$ which interchanges both the coordinates and the intrinsic degrees of freedom (i.e. the spin) of two electrons. This operator will commute with the Hamiltonian even if it contains spin forces. In the same way as it was shown for the operator $\hat{P}_{ik}$ one may show that the operator $\hat{P}'_{ik}$ has eigenvalues ± 1, and the exact stationary states of an atom must therefore also be eigenfunctions for $\hat{P}'_{ik}$. In contrast to the operator $\hat{P}_{ik}$, the operator $\hat{P}'_{ik}$ is an exact constant of the motion, and it is therefore not possible to change the symmetry property of the electron wave function in any way. Any influence which we may exert on the electrons must be independent of whether we denote a given electron by the number i or k. This means that the operator $\hat{P}'_{ik}$ commutes with any Hamiltonian whatever, and its eigenvalue is therefore a quantity which is conserved under all circumstances.

This is a general feature of the quantum mechanical description of a system of identical particles, since by the word identical we merely mean that it is under no circumstances possible to distinguish between the particles. If a wave function which describes two identical particles is symmetric or antisymmetric, this is a property of the particles which one may trace back to their creation. Creation and annihilation of particles is a problem of relativistic quantum mechanics. It has been possible to show that particles with half-integer spin, like electrons, are created with antisymmetric wave functions, while particles with integer spin are created in symmetric states.

This law of nature has a number of important consequences. A direct consequence is the Pauli exclusion principle, which states that two or more identical particles with half-integer spin cannot exist in identical states. This follows simply from the fact that the wave function for two particles in identical quantum states must necessarily be symmetric under exchange of the particles, in contradiction to the above principle.

Let us now see which influence the antisymmetrization has on the Hartree model. When one includes the electron spin in the Hartree model the wave functions are of the form

$$u = w_1(1)w_2(2)\,\ldots\,.\,w_N(N), \tag{5.5.10}$$

where

$$w_i(k) = u_{n_i l_i m_i}(\mathbf{r}_k)\chi_{m_{si}}(k). \tag{5.5.11}$$

The spatial wave function $u_{n_i l_i m_i}$ for the ith electron is here multiplied by the intrinsic wave function χ, which describes one of the two degenerate states with spin orientation $m_s = \pm\frac{1}{2}$. In (5.5.11) the index i denotes the quantum numbers, while k denotes the degrees of freedom of the kth electron. The complete antisymmetric wave function which corresponds to the Hartree wave function (5.5.10) is seen to be given by the determinant

$$u = \frac{1}{\sqrt{N!}}\begin{vmatrix} w_1(1) & w_1(2) & w_1(3) & \cdots & w_1(N) \\ w_2(1) & w_2(2) & \cdot & \cdots & w_2(N) \\ w_3(1) & w_3(2) & \cdot & \cdots & \cdot \\ \cdot & \cdot & \cdot & \cdots & \cdot \\ \cdot & \cdot & \cdot & \cdots & \cdot \\ \cdot & \cdot & \cdot & \cdots & \cdot \\ w_N(1) & w_N(2) & \cdot & \cdots & w_N(N) \end{vmatrix} \tag{5.5.12}$$

where the factor $(N!)^{-\frac{1}{2}}$ is a normalization factor. Thus if we interchange in u the spatial and intrinsic coordinates for two arbitrary electrons this corresponds to an interchange of two columns in (5.5.12) and u therefore changes sign. It is also seen directly from (5.5.12) that if all quantum numbers n, l, m and m_s are identical for two electrons then two rows are identical and u vanishes, corresponding to the fact that the state does not satisfy the Pauli principle.

It is clear that the validity of the Pauli principle does not depend on the use of the Hartree model. We discussed it in connexion with the Hartree description merely in order to have at our disposal definite examples of many-particle wave functions. We may ask now whether it makes sense at all to assume, in the derivation of the Hartree model, that the stationary states are described by product wave functions which are not properly antisymmetric. This question can be investigated by carrying out a variational calculation analogous to the one discussed in 5.4, with a trial wave function of the type (5.5.12). One then obtains a description which is called the Hartree–Fock model and which is rather more complicated than the Hartree model. The results do not usually deviate very much from the results obtained by first carrying through the ordinary Hartree calculation and then antisymmetrizing the resulting product wave function as described above.

A wave function of type (5.5.10) is the natural starting point for a Hartree approximation if the Hamilton operator contains interactions between the spin and the orbital motion. While it is necessary to include these forces if one wants to obtain an accurate description of heavy atoms, they may be disregarded in light atoms. For light atoms the wave function can then be written as a product of a spatial wave function and a spin wave function. Since the Hamilton operator here commutes with the operator $\hat{P}_{ik}$, the spatial wave function must be an eigenfunction for $\hat{P}_{ik}$.

As an example we may return to the Hartree model for the helium atom, where we may easily construct the total wave function, including spin. In order to do this we multiply the wave functions (5.5.7) and (5.5.8) by the intrinsic (spin) wave functions, $\chi(1)$ and $\chi(2)$, for the two electrons. According to the Pauli principle we must do this in such a way that the total wave function is antisymmetric under the interchange of spatial as well as intrinsic coordinates. For a state described by the spatial wave function u_+ this gives a unique answer, namely

$$u_{00} = u_+^{(1)} \frac{1}{\sqrt{2}} [\chi_{\frac{1}{2}}(1)\chi_{-\frac{1}{2}}(2) - \chi_{\frac{1}{2}}(2)\chi_{-\frac{1}{2}}(1)], \qquad (5.5.13)$$

since there only exists the quoted antisymmetric combination of the product of two spin functions $\chi_{m_1}(1)$ and $\chi_{m_2}(2)$.

The complete wave function corresponding to the state u_- is not unique. One may construct three linearly independent wave functions which are all totally antisymmetric, namely

$$u_{1,-1} = u_-^{(1)}\chi_{-\frac{1}{2}}(1)\chi_{-\frac{1}{2}}(2) \tag{5.5.14}$$

$$u_{1,0} = u_-^{(1)}\frac{1}{\sqrt{2}}[\chi_{\frac{1}{2}}(1)\chi_{-\frac{1}{2}}(2)+\chi_{\frac{1}{2}}(2)\chi_{-\frac{1}{2}}(1)] \tag{5.5.15}$$

$$u_{1,1} = u_-^{(1)}\chi_{\frac{1}{2}}(1)\chi_{\frac{1}{2}}(2). \tag{5.5.16}$$

As we shall see in the following section the wave functions (5.5.13) describe states in which the spins of the two electrons are antiparallel (singlet states), while the wave functions (5.5.14)–(5.5.16) describe states in which they are parallel (triplet states), and in which the projection of the total spin on the z-axis is $-\hbar$,0 and $\hbar$ respectively. We may thus conclude that para-helium contains the singlet states, while ortho-helium consists of triplet states.

Problems

5.5.1 In a diatomic molecule the two nuclei are identical and spinless. Show that half of the states mentioned in problem (5.1.4) are then forbidden by the Pauli principle.

How is the situation changed if the identical nuclei have spin $\frac{1}{2}$?

5.5.2 Show that the wave function (5.5.12) is normalized if the states $w_k(i)$ are orthonormal.

5.6 ADDITION OF ANGULAR MOMENTA

In the previous section it was mentioned that in the Hartree approximation one should take into account the fact that the average field from an electron in a state with $l \neq 0$ and a definite magnetic quantum number, m, is not spherically symmetric. This could give rise to great complications in the solution of the next step in the Hartree approximation, since one would have to solve the Schrödinger equation for a particle in a field without spherical symmetry. However, the situation is simplified by the fact that all the inner electrons together give rise to a spherically symmetric field, since according to the Pauli principle they usually occupy all $2l+1$ magnetic substates for a given l [cf. problem 5.4.1]. Only for the few electrons which move outside the closed shells one should take into account the fact that they interact through an asymmetric average field.

The construction of the wave function for these electrons is much simplified by the fact that the true state must be an eigenstate of the operator for the total angular momentum of the electrons. Since the total angular momentum of the electrons in the closed shells is zero [cf. problem 5.6.1], one obtains a much better approximation to the Hartree wave function if one ensures that the few outer electrons are in a state of well-defined total angular momentum. The situation is analogous to the situation we encountered in the previous section where we formed a wave function by linear combination of the degenerate states (5.5.1) and (5.5.2) which was an eigenfunction for $\hat{P}_{ik}$ which was known beforehand to be a constant of the motion.

Before we discuss the addition of angular momenta we note that the corresponding problem of addition of linear momenta is almost trivial. Thus if we know the eigenstates $u_{\mathbf{p}_i}$ of the momentum operator $\hat{\mathbf{p}}_i = -i\hbar\nabla_i$ of the ith particle, i.e.

$$\hat{\mathbf{p}}_i u_{\mathbf{p}_i}(\mathbf{r}_i) = \mathbf{p}_i u_{\mathbf{p}_i}(\mathbf{r}_i), \tag{5.6.1}$$

where $\hat{\mathbf{p}}_i = (p_{ix}, p_{iy}, p_{iz})$ are the eigenvalues, it is clear that eigenfunctions for the equation

$$\hat{\mathbf{p}}u = \mathbf{p}u, \tag{5.6.2}$$

where $\hat{\mathbf{p}}$, the operator for the total momentum

$$\hat{\mathbf{p}} = \hat{\mathbf{p}}_1 + \hat{\mathbf{p}}_2 + \cdots, \tag{5.6.3}$$

is given simply by the product wave function

$$u = u_{\mathbf{p}_1}(\mathbf{r}_1)u_{\mathbf{p}_2}(\mathbf{r}_2)\cdots . \tag{5.6.4}$$

As it is seen by applying the operator (5.6.3) to the wave function (5.6.4), the eigenvalue $\mathbf{p}$ in (5.6.2) is given by

$$\mathbf{p} = \mathbf{p}_1 + \mathbf{p}_2 + \cdots . \tag{5.6.5}$$

In the case of angular momenta the problem is more complicated because the angular momentum operators $\hat{L}_{ix}$, $\hat{L}_{iy}$, $\hat{L}_{iz}$, for the ith particle do not commute, in contrast to the linear momentum operators, $\hat{p}_{ix}$, p_{iy}, p_{iz}. We can therefore only specify $(\hat{\mathbf{L}}_i)^2$ and $\hat{L}_{iz}$ and the angular momentum eigenstates for the ith particle corresponding to (5.6.1) are given by

$$|\hat{\mathbf{L}}_i|^2 u_{l_i m_i}(\mathbf{r}_i) = \hbar^2 l_i(l_i+1)u_{l_i m_i}(\mathbf{r}_i) \tag{5.6.6}$$

and

$$\hat{L}_{iz} u_{l_i m_i}(\mathbf{r}_i) = \hbar m_i u_{l_i m_i}(\mathbf{r}_i). \tag{5.6.7}$$

Now it is easy to construct an eigenfunction u for the operator $\hat{J}_z$ for the z-component of the total angular momentum:

$$\hat{J}_z u = \hbar m u, \tag{5.6.8}$$

where

$$\hat{J}_z = \hat{L}_{1z}+\hat{L}_{2z}+\cdots. \tag{5.6.9}$$

One finds as in the case of linear momenta that

$$u = u_{l_1 m_1}(\mathbf{r}_1)u_{l_2 m_2}(\mathbf{r}_2)\cdots \tag{5.6.10}$$

satisfies (5.6.8) with

$$m = m_1+m_2+\cdots\cdots. \tag{5.6.11}$$

The wave function (5.6.10) is, however, in general not an eigenfunction for the magnitude of the total angular momentum

$$|\hat{\mathbf{J}}|^2 = |\hat{\mathbf{L}}_1+\hat{\mathbf{L}}_2+\ldots.|^2, \tag{5.6.12}$$

as is obvious from the fact that the x- and y-components of the angular momenta are not welldefined.

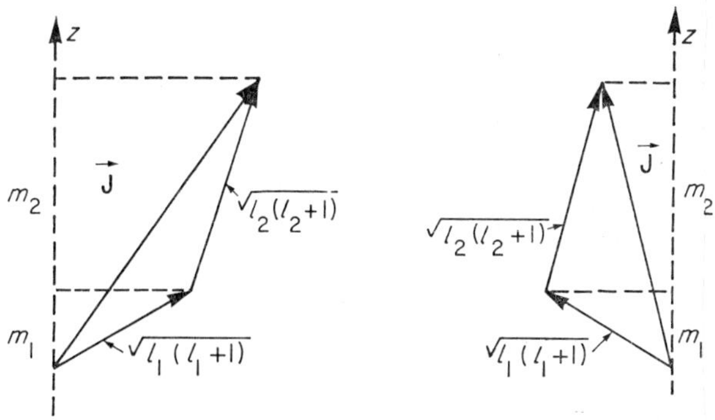

Fig. 5.1

By forming a linear combination of the degenerate states (5.6.10) corresponding to different choices of m_1, m_2, etc., one may, as we shall see now, construct a state u_{jm} which is both an eigenfunction for $\hat{J}_z$ and for $|\hat{\mathbf{J}}|^2$, i.e. a state which satisfies both (5.6.8) and

$$|\hat{\mathbf{J}}|^2 u_{jm} = \hbar^2 j(j+1)u_{jm}. \tag{5.6.13}$$

Let us confine the discussion to two particles, where we consequently assume

$$u_{jm} = \sum_{m_1 m_2} C^{jm}_{l_1 m_1 l_2 m_2} u_{l_1 m_1} u_{l_2 m_2}. \tag{5.6.14}$$

The coefficients C, which we are going to determine, may depend on all the indices indicated. Inserting (5.6.14) in (5.6.8) we note first that u_{jm} is only an eigenfunction of $\hat{J}_z$ if

$$C^{jm}_{l_1 m_1 l_2 m_2} = 0 \quad \text{for} \quad m_1 + m_2 \neq m.$$

We now insert (5.6.14) in (5.6.13) writing $|\hat{\mathbf{J}}|^2$ in the form

$$\begin{aligned} |\hat{\mathbf{J}}|^2 &= |\hat{\mathbf{L}}_1|^2 + |\hat{\mathbf{L}}_2|^2 + 2\hat{\mathbf{L}}_1 \cdot \hat{\mathbf{L}}_2 \\ &= |\hat{\mathbf{L}}_1|^2 + |\hat{\mathbf{L}}_2|^2 + 2\hat{L}_{1z}\hat{L}_{2z} + \hat{L}_{1+}\hat{L}_{2-} + \hat{L}_{1-}\hat{L}_{2+}, \end{aligned} \tag{5.6.15}$$

where the $\pm$ components are defined by the equations (5.2.12) and (5.2.14). One finds then

$$\begin{aligned} \sum_{m_1 m_2} C^{jm}_{l_1 m_1 l_2 m_2} \{[l_1(l_1+1) + l_2(l_2+1) - j(j+1) + 2m_1 m_2]\hbar^2 u_{l_1 m_1} u_{l_2 m_2} \\ + A(l_1 m_1) B(l_2 m_2) u_{l_1 m_1+1} u_{l_2 m_2-1} \\ + B(l_1 m_1) A(l_2 m_2) u_{l_1 m_1-1} u_{l_2 m_2+1}]\} = 0, \end{aligned} \tag{5.6.16}$$

where we have used the relations (5.2.16) and (5.2.17). Introducing the expressions (5.2.24) and (5.2.22), redefining the indices of the last two terms, one obtains

$$\begin{aligned} \sum_{m_1 m_2} \{C^{jm}_{l_1 m_1 l_2 m_2} [l_1(l_1+1) + l_2(l_2+1) - j(j+1) + 2m_1 m_2] \\ + C^{jm}_{l_1 m_1 - 1\; l_2 m_2 + 1} [(l_1+m_1)(l_1-m_1+1)(l_2+m_2+1)(l_2-m_2)]^{\frac{1}{2}} \\ + C_{l_1 m_1 + 1\; l_2 m_2 - 1} [(l_1+m_1+1)(l_1-m_1)(l_2+m_2)(l_2-m_2+1)]^{\frac{1}{2}}\} \\ u_{l_1 m_1} u_{l_2 m_2} = 0. \end{aligned} \tag{5.6.17}$$

The condition for (5.6.14) to be an eigenfunction for $|\hat{\mathbf{J}}|^2$ is now apparently that all terms in (5.6.17) vanish. The set of homogenous, linear equations in $C^{jm}_{l_1 m_1 l_2 m_2}$ which results from this condition have solutions only for such values of j that the equations become linearly dependent. The equations (5.6.17) thus determine both the possible values of j and the coefficients C, apart from a common normalization. We shall not attempt here to carry through such a general calculation, but only mention the result. For given values of l_1 and l_2 one finds that j can take on the values

$$j = |l_1 - l_2|, |l_1 - l_2| + 1, \ldots\ldots, l_1 + l_2 - 1, l_1 + l_2, \quad (5.6.18)$$

where the lowest value corresponds to the case that the angular momenta $\hat{\mathbf{L}}_1$ and $\hat{\mathbf{L}}_2$ are antiparallel, while the largest value corresponds to the case that they are parallel. The coefficients C are denoted vector-coupling coefficients or Clebsch–Gordan coefficients. One may derive a complicated explicit formula for C. For practical use extensive tables have been published.

In this section we have considered the addition of angular momenta arising from the orbital motion of the particles. In many situations, especially if one considers spin-orbit forces, it is important to be able to construct states in which the sum of orbital and spin angular momenta have a definite value. For a single particle with spin such states are of the form

$$u_{jm} = \sum_{m_1 m_s} C^{jm}_{l_1 m_1 s m_s} u_{l_1 m_1} \chi_{s m_s}, \quad (5.6.19)$$

where u_{jm} must satisfy (5.6.8) and (5.6.13) with $\hat{\mathbf{S}}$ substituted for $\hat{\mathbf{L}}_2$. Since, however, in the above calculations we only used the general rules for angular momenta which were derived in 5.2, we may directly apply the result (5.6.17) allowing the quantum numbers to take on half-integer values.

For the case $l_2 = \frac{1}{2}$, (5.6.17) leads to two homogenous equations with two unknowns which, in accordance with (5.6.18), can be solved for

$$j = \begin{cases} l_1 - \frac{1}{2} \\ l_1 + \frac{1}{2}. \end{cases}$$

If one demands furthermore that the state u_{jm} be normalized, one may easily calculate the Clebsch–Gordan coefficients. The result is given in Table 5.2.

One may of course also add the intrinsic angular momenta for two particles and thereby construct a wave function for which the total spin has a definite value. As an example we consider the addition of two angular momenta of magnitude $\frac{1}{2}$. One finds here four possible states, one with $j = 0$ (and $m = 0$) and three with $j = 1$ ($m = -1$, 0, 1). Evaluating the vector-coupling coefficients by means of

Table 5.2, one sees that these states are merely the ones encountered in the spin part of the wave functions (5.5.13)–(5.5.16).

For systems of more than two particles one may construct wave functions of well-defined total angular momentum by adding first the angular momenta of two of the particles, then adding to this total angular momentum the angular momentum of the third particle etc.

Table 5.2

	$C^{jm}_{j_1 m_1 \frac{1}{2} m_2}$	
	$m_2 = \frac{1}{2}$	$m_2 = -\frac{1}{2}$
$j = j_1 + \frac{1}{2}$	$\sqrt{\dfrac{j_1+m+\frac{1}{2}}{2j_1+1}}$	$\sqrt{\dfrac{j_1-m+\frac{1}{2}}{2j_1+1}}$
$j = j_1 - \frac{1}{2}$	$-\sqrt{\dfrac{j_1-m+\frac{1}{2}}{2j_1+1}}$	$\sqrt{\dfrac{j_1+m+\frac{1}{2}}{2j_1+1}}$

Problem

5.6.1 Show that the antisymmetrized product wave function describing $2l+1$ particles filling the $2l+1$ magnetic substates for a given angular momentum l, describes a state of total angular momentum zero. To prove this one may, for example, evaluate the result of applying $\hat{J}_z$, $\hat{J}_+$ and $\hat{J}_-$ to the wave function.

Index